Fundamentals of Gas Dynamics

T0172263

V. Babu

Fundamentals of Gas Dynamics

Second Edition

V. Babu
Department of Mechanical Engineering
Indian Institute of Technology Madras
Chennai, Tamil Nadu, India

ISBN 978-3-030-60821-7   ISBN 978-3-030-60819-4   (eBook)
https://doi.org/10.1007/978-3-030-60819-4

Jointly published with ANE Books India
In addition to this printed edition, there is a local printed edition of this work available via Ane Books in
South Asia (India, Pakistan, Sri Lanka, Bangladesh, Nepal and Bhutan) and Africa (all countries in the
African subcontinent).
ISBN of the Co-Publisher's edition: 9781910390009

Cover illustration: Schlieren picture of an under-expanded flow issuing from a convergent divergent
nozzle. Prandtl Meyer expansion waves in the divergent portion as the flow goes around the convex
throat can be seen. Expansion fans, reflected oblique shocks and the alternate swelling and compression
of the jet are clearly visible. Courtesy: P. K. Shijin, Ph.D. scholar, Department of Mechanical
Engineering, IIT Madras.

This Springer imprint is published by the registered company Springer Nature Switzerland AG
The registered company address is: Gewerbestrasse 11, 6330 Cham, Switzerland

Dedicated to my wife Chitra and son Aravindh for their enduring patience and love

Preface

I am happy to introduce this edition of the book *Fundamentals of Gas Dynamics*. Readers of the first edition should be able to see changes in all the chapters—changes in the development of the material, new materials, and figures as well as more end-of-chapter problems. In keeping with the spirit of the first edition, the additional exercise problems are drawn from practical applications to enable the student to make the connection from concept to application.

Owing to the ubiquitous nature of steam power plants around the world, it is important for mechanical engineering students to learn the gas dynamics of steam. With this in mind, a new chapter on the gas dynamics of steam has been added in this edition. This is somewhat unusual since this topic is usually introduced in textbooks on steam turbines and not in gas dynamics texts. In my opinion, introducing this in a gas dynamics text is logical and in fact makes it easy for the students to learn the concepts. In developing this material, I have assumed that the reader would have gone through a fundamental course in thermodynamics and so would be familiar with calculations involving steam. Steam tables for use in these calculations have also been added at the end of the book. I would like to thank Prof. Korpela of Ohio State University for generating these tables and allowing me to include them in the book.

I wish to thank the readers, who purchased the first edition and gave me many suggestions as well as for pointing out errors. To the extent possible, the errors have been corrected and the suggestions have been incorporated in this edition. If there are any errors or if you have any suggestions for improving the exposition of any topic, please feel free to communicate them to me via e-mail (vbabu@iitm.ac.in). I would like to take this opportunity to thank Prof. S. R. Chakravarthy of IIT Madras for his suggestion concerning the definition of compressibility. I have taken this further and connected it with Rayleigh flow in the incompressible limit. The effect of different γ on the property changes across a normal shock wave is now included in Chap. 3. The development of the process curve in Chaps. 4 and 5 has been done by directly relating the changes in properties to changes in stagnation temperature and entropy, respectively. In Chap. 6, I have added a figure showing the variation of static pressure along a CD nozzle as well as the variation of exit static pressure to the ambient pressure. Hopefully, this will make it easier for the students to understand over- and under-expanded flow.

Once again I would like to express my heartfelt gratitude to my teachers, who taught me so much without expecting anything in return. I can only hope that I succeed in giving back at least a fraction of the knowledge and wisdom that I received from them. My advisor, mentor, and friend, Prof. Seppo Korpela has been an inspiration to me and his constant and patient counsel has helped me enormously. I am indebted to my parents for the sacrifices they made to impart a good education to me. This is not a debt that can be repaid. But for the constant support and encouragement from my wife and son, this edition and the other books that I have written would not have been possible.

Finally, I would like to thank my former students P. S. Tide, S. Somasundaram, and Anandraj Hariharan for diligently working out the examples and exercise problems and my current student P. K. Shijin for carefully proofreading the manuscript and making helpful suggestions. Thanks are due in addition to Prof. P. S. Tide for preparing the Solutions Manual.

Chennai, India   V. Babu
April 2020

Contents

About the Author

Dr. V. Babu is currently Professor in the Department of Mechanical Engineering at IIT Madras, Chennai. He received his B.E. in Mechanical Engineering from REC Trichy in 1985 and Ph.D. from Ohio State University in 1991. He worked as a Post-Doctoral researcher at the Ohio State University from 1991 to 1995. He was a Technical Specialist at the Ford Scientific Research Lab, Michigan from 1995 to 1998. He received the Henry Ford Technology Award in 1998 for the development and deployment of a virtual wind tunnel. He has four U.S. patents to his credit. He has published technical papers on simulations of fluid flows including plasmas and non-equilibrium flows, computational aerodynamics and aeroacoustics, scientific computing and ramjet, scramjet engines. His primary research specialization is CFD and he is currently involved in the simulation of high speed reacting flows, prediction of jet noise, simulation of fluid flows using the lattice Boltzmann method and high performance computing.

Chapter 1
Introduction

Compressible flows are encountered in many applications in aerospace and mechanical engineering. Some examples are flows in nozzles, compressors, turbines, and diffusers. In aerospace engineering, in addition to these examples, compressible flows are seen in external aerodynamics, aircraft, and rocket engines. In almost all of these applications, air (or some other gas or mixture of gases) is the working fluid. However, steam can be the working substance in turbomachinery applications. Thus, the range of engineering applications in which compressible flow occurs is quite large and hence a clear understanding of the dynamics of compressible flow is essential for engineers.

1.1 Compressibility of Fluids

All fluids are compressible to some extent or other. The compressibility of a fluid is defined as

$$\tau = -\frac{1}{v}\frac{\partial v}{\partial P} \tag{1.1}$$

where v is the specific volume and P is the pressure. The change in specific volume corresponding to a given change in pressure, will, of course, depend upon the compression process. That is, for a given change in pressure, the change in specific volume will be different between an isothermal and an adiabatic compression process.

The definition of compressibility actually comes from thermodynamics. Since the specific volume $v = v(T, P)$, we can write

$$dv = \left(\frac{\partial v}{\partial P}\right)_T dP + \left(\frac{\partial v}{\partial T}\right)_P dT$$

© The Author(s) 2021
V. Babu, *Fundamentals of Gas Dynamics*,
https://doi.org/10.1007/978-3-030-60819-4_1

From the first term, we can define the isothermal compressibility as $-\frac{1}{v}\left(\frac{\partial v}{\partial P}\right)_T$ and, from the second term, we can define the coefficient of volume expansion as $\frac{1}{v}\left(\frac{\partial v}{\partial T}\right)_P$. The second term represents the change in specific volume (or equivalently density) due to a change in temperature. For example, when a gas is heated at constant pressure, the density decreases and the specific volume increases. This change can be large, as is the case in most combustion equipment, without necessarily having any implications on the compressibility of the fluid. It thus follows that compressibility effect is important only when the change in specific volume (or equivalently density) is due largely to a change in pressure.

If the above equation is written in terms of the density ρ, we get

$$\tau = \frac{1}{\rho}\frac{\partial \rho}{\partial P} \tag{1.2}$$

The isothermal compressibility of water and air under standard atmospheric conditions are 5×10^{-10} m^2/N and 10^{-5} m^2/N. Thus, water (in liquid phase) can be treated as an incompressible fluid in all applications. On the contrary, it would seem that air, with a compressibility that is five orders of magnitude higher, has to be treated as a compressible fluid in all applications. Fortunately, this is not true when the flow is involved.

1.2   Compressible and Incompressible Flows

It is well known from high school physics that sound (pressure waves) propagates in any medium with a speed that depends on the bulk compressibility. The less compressible the medium, the higher the speed of sound. Thus, speed of sound is a convenient reference speed, when flow is involved. Speed of sound in air under normal atmospheric conditions is 330 m/s. The implications of this when there is flow are as follows. Let us say that we are considering the flow of air around an automobile traveling at 120 kph (about 33 m/s). This speed is 1/10th of the speed of sound. In other words, compared with 120 kph, sound waves travel 10 times faster. Since the speed of sound *appears* to be high compared with the highest velocity in the flow field, the medium behaves as though it were incompressible. As the flow velocity becomes comparable to the speed of sound, compressibility effects become more prominent. In reality, the speed of sound itself can vary from one point to another in the flow field and so the velocity at each point has to be compared with the speed of sound at that point. This ratio is called the Mach number, after Ernst Mach who made pioneering contributions in the study of the propagation of sound waves. Thus, the Mach number at a point in the flow can be written as

$$M = \frac{u}{a} \tag{1.3}$$

where u is the velocity magnitude at any point and a is the speed of sound at that point.

We can come up with a quantitative criterion to give us an idea about the importance of compressibility effects in the flow by using simple scaling arguments as follows. From Bernoulli's equation for steady flow, it follows that $\Delta P \sim \rho U^2$, where U is the characteristic speed. It will be shown in the next chapter that the speed of sound $a = \sqrt{\Delta P / \Delta \rho}$, wherein ΔP and $\Delta \rho$ correspond to an isentropic process. Thus,

$$\frac{\Delta \rho}{\rho} = \frac{1}{\rho} \frac{\Delta \rho}{\Delta P} \Delta P = \frac{U^2}{a^2} = M^2 \tag{1.4}$$

On the other hand, upon rewriting Eq. 1.2 for an isentropic process, we get

$$\frac{\Delta \rho}{\rho} = \tau_{isentropic} \Delta P$$

Comparison of these two equations shows clearly that, in the presence of a flow, density changes are proportional to the square of the Mach number.[1] It is customary to assume that the flow is essentially incompressible if the change in density is less than 10% of the mean value.[2] It thus follows that compressibility effects are significant only when the Mach number exceeds 0.3.

1.3 Perfect Gas Equation of State

In this text, we assume throughout that air behaves as a perfect gas. The equation of state can be written as

$$Pv = RT \tag{1.5}$$

where T is the temperature.[3] R is the particular gas constant and is equal to $\mathcal{R}/M$ where $\mathcal{R} = 8314$ J/kmol/K is the Universal Gas Constant and M is the molecular weight in units of $kg/kmol$. Equation 1.5 can be written in many different forms depending upon the application under consideration. A few of these forms are presented here for the sake of completeness. Since the specific volume $v = 1/\rho$, we can write

$$P = \rho RT$$

or, alternatively, as

---

[1] This is true for steady flows only. For unsteady flows, density changes are proportional to the Mach number.

[2] Provided the change is predominantly due to a change in pressure.

[3] In later chapters, this will be referred to as the static temperature.

$$PV = mRT$$

where m is the mass and V is the volume. If we define the concentration c as $(m/M)(1/V)$, then,

$$P = cRT \tag{1.6}$$

Here c has units of kmol/m^3. The mass density ρ can be related to the particle density n (particles/m^3) through the relationship $\rho = nM/N_A$. Here we have used the fact that 1 *kmol* of any substance contains Avogadro number of molecules ($N_A = 6.023 \times 10^{26}$). Thus

$$P = n\frac{R}{N_A}T = nk_BT \tag{1.7}$$

where k_B is the Boltzmann constant.

1.3.1 Continuum Hypothesis

In our discussion so far, we have tacitly assumed that properties such as pressure, density, velocity , and so on can be evaluated without any ambiguity. While this is intuitively correct, it deserves a closer examination.

Consider the following thought experiment. A cubical vessel of a side dimension L contains a certain amount of a gas. One of the walls of the vessel has a viewport to allow observations of the contents within a fixed observation volume. We now propose to measure the density of the gas at an instant as follows—count the number of molecules within the observation volume; multiply this by the mass of each molecule and then divide by the observation volume.

To begin with, let there be 100 molecules inside the vessel. We would notice that the density values measured in the aforementioned manner fluctuate wildly going down even to zero at some instants. If we increase the number of molecules progressively to 10^3, 10^4, 10^5, and so on, we would notice that the fluctuations begin to diminish and eventually die out altogether. Increasing the number of molecules beyond this limit would not change the measured value for the density.

We can carry out another experiment in which we attempt to measure the pressure using a pressure sensor mounted on one of the walls. Since the pressure exerted by the gas is the result of the collisions of the molecules on the walls, we would notice the same trend as we did with the density measurement. That is, the pressure measurements too exhibit fluctuations when there are few molecules and the fluctuations die out with an increasing number of molecules. The measured value, once again, does not change when the number of molecules is increased beyond a certain limit.

We can intuitively understand that, in both these experiments, when the number of molecules is less, the molecules travel freely for a considerable distance before encountering another molecule or a wall. As the number of molecules is increased, the distance that a molecule on an average can travel between collisions (which is termed as the mean free path, denoted usually by λ) decreases as the collision frequency increases. Once the mean free path decreases below a limiting value, measured property values do not change anymore. The gas is then said to behave as a *continuum*. The determination of whether the actual value for the mean free path is small or not has to be made relative to the physical dimensions of the vessel. For instance, if the vessel is itself only about 1 μm in dimension on each side, then a mean free path of 1 μm is not at all small! Accordingly, a parameter known as the Knudsen number (Kn) which is defined as the ratio of the mean free path (λ) to the characteristic dimension (L) is customarily used. Continuum is said to prevail when $Kn \ll 1$. In reality, once the Knudsen number exceeds 10^{-2} or so, the molecules of the gas cease to behave as a continuum.

It is well known from kinetic theory of gases that the mean free path is given as

$$\lambda = \frac{1}{\sqrt{2}\pi d^2 n} \tag{1.8}$$

where d is the diameter of the molecule and n is the number density.

Example 1.1 Determine whether continuum prevails in the following two practical situations: (a) an aircraft flying at an altitude of 10 km where the ambient pressure and temperature are 26.5 kPa and 230 K respectively, and (b) a hypersonic cruise vehicle flying at an altitude of 32 km where the ambient pressure and temperature are 830 Pa and 230 K respectively. Take $d = 3.57 \times 10^{-10}$ m.

Solution In both the cases, it is reasonable to assume the characteristic dimension L to be 1 m.

(a) Upon substituting the given values of the ambient pressure and temperature into the equation of state, $P = nk_BT$, we get $n = 8.34 \times 10^{24}$ particles/m^3. Hence

$$\lambda = \frac{1}{\sqrt{2}\pi d^2 n} = 2.12 \times 10^{-7} \text{ m}$$

Therefore, the Knudsen number $Kn = \lambda/L = 2.12 \times 10^{-7}$.

(b) Following the same procedure as before, we can easily obtain $Kn = 6.5 \times 10^{-6}$.

It is thus clear that, in both cases, it is quite reasonable to assume that continuum prevails.

1.4 Calorically Perfect Gas

In the study of compressible flows, we need, in addition to the equation of state, an equation relating the internal energy to other measurable properties. The internal energy, strictly speaking, is a function of two thermodynamic properties, namely, temperature and pressure. In reality, the dependence on pressure is very weak for gases and hence is usually neglected. Such gases are called *thermally perfect* and for them $e = f(T)$. The exact nature of this function is examined next.

From a molecular perspective, it can be seen intuitively that the internal energy will depend on the number of modes in which energy can be stored (also known as degrees of freedom) by the molecules (or atoms) and the amount of energy that can be stored in each mode. For monatomic gases, the atoms have the freedom to move (and hence store energy in the form of kinetic energy) in any of the three coordinate directions.

For diatomic gases, assuming that the molecules can be modeled as "dumb bells", additional degrees of freedom are possible. These molecules, in addition to translational motion along the three axes, can also rotate about these axes. Hence, energy storage in the form of rotational kinetic energy is also possible. In reality, since the moment of inertia about the "dumb bell" axis is very small, the amount of kinetic energy that can be stored through rotation about this axis is negligible. Thus, rotation adds essentially two degrees of freedom only. In the "dumb bell" model, the bonds connecting the two atoms are idealized as springs. When the temperature increases beyond 600 K or so, these springs begin to vibrate and so energy can now be stored in the form of vibrational kinetic energy of these springs. When the temperature becomes high (>2000 K), transition to other electronic levels and dissociation take place and at even higher temperatures the atoms begin to ionize. These effects do not represent degrees of freedom.

Having identified the number of modes of energy storage, we now turn to the amount of energy that can be stored in each mode. The classical equipartition energy principle states that each degree of freedom, when "fully excited", contributes $1/2\, RT$ to the internal energy per unit mass of the gas. The term "fully excited" means that no more energy can be stored in these modes. For example, the translational mode becomes fully excited at temperatures as low as 3 K itself. For diatomic gases, the rotational mode is fully excited beyond 600 K and the vibrational mode beyond 2000 K or so. Strictly speaking, all the modes are quantized and so the energy stored in each mode has to be calculated using quantum mechanics. However, the spacing between the energy levels for the translational and rotational modes is small enough, we can assume equipartition principle to hold for these modes.

We can thus write

$$e = \frac{3}{2}RT$$

for monatomic gases and

$$e = \frac{3}{2}RT + RT + \frac{h\nu/k_B T}{e^{h\nu/k_B T} - 1}RT$$

for diatomic gases. In the above expression, ν is the fundamental vibrational frequency of the molecule. Note that for large values of T, the last term approaches RT. We have not derived this term formally as it would be well outside the scope of this book. Interested readers may see the book by Anderson for full details.

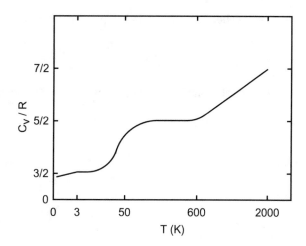

Fig. 1.1 Variation of C_v/R with temperature for diatomic gases

The enthalpy per unit mass can now be calculated by using the fact that

$$h = e + Pv = e + RT$$

We can calculate C_v and C_p from these equations by using the fact that $C_v = \partial e/\partial T$ and $C_p = \partial h/\partial T$. Thus

$$C_v = \frac{3}{2}R$$

for monatomic gases and

$$C_v = \frac{5}{2}R + \frac{(h\nu/k_BT)^2\, e^{h\nu/k_BT}}{\left(e^{h\nu/k_BT} - 1\right)^2}R$$

for diatomic gases. The variation of C_v/R is illustrated schematically in Fig. 1.1. It is clear from this figure that $C_v = 5/2R$ in the temperature range $50K \leq T \leq 600$ K. In this range, $C_p = 7/2R$, and thus the ratio of specific heats $\gamma = 7/5$ for diatomic gases. For monatomic gases, it is easy to show that $\gamma = 5/3$. In this temperature range, where C_v and C_p are constants, the gases are said to be *calorically perfect*. We will assume calorically perfect behavior in all the subsequent chapters.[4] Also, for a calorically perfect gas, since $h = C_pT$ and $e = C_vT$, it follows from the definition of enthalpy that

$$C_p - C_v = R \qquad (1.9)$$

---

[4]In all the worked examples (except those in the last chapter), we have taken air to be the working fluid. It is assumed to be calorically perfect with molecular weight 28.8 kg/kmol and $\gamma = 1.4$.

This is called Meyer's relationship. In addition, it is easy to see that

$$C_v = \frac{R}{\gamma - 1} \, , \qquad C_p = \frac{\gamma R}{\gamma - 1} \tag{1.10}$$

These relationships will be used extensively throughout the following chapters.

Chapter 2
One-Dimensional Flows—Basics

In this chapter, we discuss some fundamental concepts in the study of compressible flows. Throughout this book, we assume the flow to be one-dimensional or quasi-one-dimensional. A flow is said to be one-dimensional, if the flow properties change only along the flow direction. The fluid can have velocity either along the flow direction or both along and perpendicular to it. Oblique shock waves and Prandtl Meyer expansion/compression waves discussed in later chapters are examples of the latter. We begin with a discussion of one-dimensional flows which belong to the former category i.e., with velocity along the flow direction only.

2.1 Governing Equations

The governing equations for frictionless, adiabatic, steady, one-dimensional flow of a calorically perfect gas can be written in differential form as

$$d(\rho u) = 0 \tag{2.1}$$

$$dP + \rho u du = 0 \tag{2.2}$$

and

$$dh + d\left(\frac{u^2}{2}\right) = 0 \tag{2.3}$$

These equations express mass, momentum, and energy conservation, respectively. In addition, changes in flow properties must also obey the second law of thermodynamics. Thus,

$$ds \geq \left(\frac{\delta q}{T}\right)_{rev} \tag{2.4}$$

© The Author(s) 2021
V. Babu, *Fundamentals of Gas Dynamics*,
https://doi.org/10.1007/978-3-030-60819-4_2

where s is the entropy per unit mass and q is the heat interaction, also expressed on a per unit mass basis. The subscript refers to a reversible process. From the first law of thermodynamics, we have

$$de = C_v dT = \delta q_{rev} - P dv \tag{2.5}$$

Since $\delta q_{rev} = T ds$ from Eq. 2.4, and using the equation of state $Pv = RT$, and its differential form $P dv + v dP = R dT$, we can write

$$ds = C_v \frac{dT}{T} + R \frac{dv}{v} = C_v \frac{dP}{P} + C_p \frac{dv}{v} = C_p \frac{dT}{T} - R \frac{dP}{P} \tag{2.6}$$

Note that Eq. 2.2 is written in the so-called nonconservative form. By using Eq. 2.1, we can rewrite Eq. 2.2 in conservative form as follows.

$$dP + d\left(\rho u^2\right) = 0 \tag{2.7}$$

Equations 2.1, 2.7, 2.3 and 2.4 can be integrated between any two points in the flow field to give

$$\rho_1 u_1 = \rho_2 u_2 \tag{2.8}$$

$$P_1 + \rho_1 u_1^2 = P_2 + \rho_2 u_2^2 \tag{2.9}$$

$$h_1 + \frac{u_1^2}{2} = h_2 + \frac{u_2^2}{2} \tag{2.10}$$

and

$$s_2 - s_1 = \int_1^2 \frac{\delta q}{T} + \sigma_{irr} \tag{2.11}$$

Here, σ_{irr} represents entropy generated due to irreversibilities. It is equal to zero for an isentropic flow and is greater than zero for all other flows. It follows then from Eq. 2.11 that entropy change during an adiabatic process must increase or remain the same. The latter process, which is adiabatic and reversible is known as an isentropic process. It is important to realize that while all adiabatic and reversible processes are isentropic, the converse need not be true. This can be seen from Eq. 2.11, since with the removal of an appropriate amount of heat, the entropy increase due to irreversibilities can be offset entirely (at least in principle), thereby rendering an irreversible process isentropic. Equation 2.11 is not in a convenient form for evaluating entropy change during a process. For this purpose, we can integrate Eq. 2.6 from the initial to the final state during the process. This gives

$$s_2 - s_1 = C_v \ln \frac{T_2}{T_1} + R \ln \frac{v_2}{v_1}$$

$$= C_v \ln \frac{P_2}{P_1} + C_p \ln \frac{v_2}{v_1} \qquad (2.12)$$

$$= C_p \ln \frac{T_2}{T_1} - R \ln \frac{P_2}{P_1}$$

The flow area does not appear in any of the above equations as they stand. When we discuss one-dimensional flow in ducts and passages, this can be introduced quite easily. Also, it is important to keep in mind that, when points 1 and 2 are located across a wave (say, a sound wave or shock wave), the derivatives of the flow properties will be discontinuous.

2.2 Acoustic Wave Propagation Speed

Equations 2.8, 2.9, 2.10, and 2.12 admit different solutions, which we will see in the subsequent chapters. The most basic solution is the expression for the speed of sound, which we will derive in this section.

Consider an acoustic wave propagating into quiescent air as shown in Fig. 2.1. Although the wavefront is spherical, at any point on the wavefront, the flow is essentially one-dimensional as the radius of curvature of the wavefront is large when compared with the distance across which the flow properties change. If we switch to a reference frame in which the wave appears stationary, then the flow approaches the wave with a velocity equal to the wave speed in the stationary frame of reference and moves away from the wave with a slightly different velocity. As a result of going through the acoustic wave, the flow properties change by an infinitesimal amount and the process is isentropic. Thus, we can take $u_2 = u_1 + du_1$, $P_2 = P_1 + dP_1$, and $\rho_2 = \rho_1 + d\rho_1$. Substitution of these into Eqs. 2.8 and 2.9 gives

$$\rho_1 u_1 = (\rho_1 + d\rho_1)(u_1 + du_1)$$

and

$$P_1 + \rho_1 u_1^2 = P_1 + dP_1 + (\rho_1 + d\rho_1)(u_1 + du_1)^2$$

If we neglect the product of differential terms, then we can write

$$\rho_1 du_1 + u_1 d\rho_1 = 0$$

and

$$dP_1 + 2\rho_1 u_1 du_1 + u_1^2 d\rho_1 = 0$$

Fig. 2.1 Propagation of a
sound wave into a quiescent
fluid

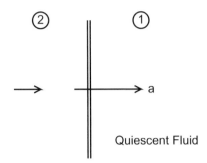

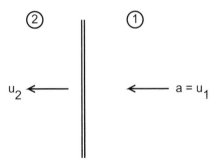

Upon combining these two equations, we get

$$\frac{dP_1}{d\rho_1} = u_1^2$$

As mentioned earlier, u_1 is equal to the speed of sound a and so

$$a = \sqrt{\left(\frac{dP}{d\rho}\right)_s} \qquad (2.13)$$

where the subscript 1 has been dropped for convenience. Furthermore, we have also
explicitly indicated that the process is isentropic. Since the process is isentropic,
$ds = 0$, and so from Eq. 2.6,

$$C_v \frac{dP}{P} + C_p \frac{dv}{v} = 0$$

Since $\rho = 1/v$, $dv/v = -d\rho/\rho$ and so

$$\frac{dP}{d\rho} = \frac{C_p}{C_v} \frac{P}{\rho} = \gamma RT$$

Thus,

$$a = \sqrt{\gamma RT} \qquad (2.14)$$

This expression is valid for a non-reacting mixture of ideal gases as well, with the understanding that γ is the ratio of specific heats for the mixture and R is the particular gas constant for the mixture.[1]

2.2.1 Mach Number

The Mach number has already been defined in Eq. 1.3 and we are now in a position to take a closer look at it. Since it is defined as a ratio, changes in the Mach number are the outcome of either changes in velocity, speed of sound, or both. Speed of sound itself varies from point to point and is proportional to the square root of the temperature as seen from Eq. 2.14. Thus, any deductions of the velocity or temperature variation from a given variation of Mach number cannot be made in a straightforward manner. For example, the velocity at the entry to the combustor in an aircraft gas turbine engine may be as high as 200 m/s, but the Mach number is usually 0.3 or less due to the high static temperature of the fluid.

2.3 Reference States

In the study of compressible flows and indeed in fluid mechanics, it is conventional to define certain reference states. These allow the governing equations to be simplified and written in dimensionless form so that the important parameters can be identified.

---

[1]Equation 2.14 is not valid for a reacting flow, since chemical reactions are by nature irreversible and hence the process across the sound wave cannot be isentropic. However, two limiting conditions can be envisaged and the speed of sound corresponding to these conditions can still be evaluated using Eq. 2.14. These are the frozen and equilibrium conditions. In the former case, the reactions are assumed to be frozen and hence the mixture is essentially non-reacting. The speed of sound for this mixture can be calculated using Eq. 2.14 appropriately. In the latter case, reactions are still taking place but the mixture is at chemical equilibrium and hence $ds = 0$. Once the equilibrium composition and temperature are known, the speed of sound for the equilibrium mixture can be determined, again using Eq. 2.14. In reality, the reactions neither have to be frozen nor do they have to be at equilibrium. These are simply two limiting situations, which allow us to get a bound of the speed of sound for the actual case.

In the context of compressible flows, the solution procedure can also be made simpler and in addition the important physics in the flow can be brought out clearly by the use of these reference states. Two such reference states are discussed next.

2.3.1 Sonic State

Since the speed of sound plays a crucial role in compressible flows, it is convenient to use the sonic state as a reference state. The sonic state is the state of the fluid at that point in the flow field where the velocity is equal to the speed of sound. Properties at the sonic state are usually denoted with a * *viz.,* P^*, T^*, ρ^*, and so on. Of course, u^* itself is equal to a and so the Mach number $M = 1$ at the sonic state. The sonic reference state can be thought of as a global reference state since it is attained only at one or a few points in the flow field. For example, in the case of choked isentropic flow through a nozzle, the sonic state is achieved in the throat section. In some other cases, such as flow with heat addition or flow with friction, the sonic state may not even be attained anywhere in the actual flow field, but is still defined in a hypothetical sense and is useful for analysis. The importance of the sonic state lies in the fact that it separates subsonic ($M < 1$) and supersonic ($M > 1$) regions of the flow. Since information travels in a compressible medium through acoustic waves, the sonic state separates regions of flow that are fully accessible (subsonic) and those that are not (supersonic).

Note that the dimensionless velocity u/u^* at a point is **not** equal to the Mach number at that point since u^* is not the speed of sound at that point.[2]

2.3.2 Stagnation State

Let us consider a point in a one-dimensional flow and assume that the state at this point is completely known. This means that the pressure, temperature, and velocity at this point are known. We now carry out a thought experiment in which an isentropic, deceleration process takes the fluid from the present state to one with zero velocity. The resulting end state is called the stagnation state corresponding to the known initial state. Thus, the stagnation state at a point in the flow field is defined as the thermodynamic state that would be reached from the given state at that point, at the end of an isentropic, deceleration process to zero velocity. Note that the stagnation state is a local state contrary to the sonic state. Hence, the stagnation state can change from one point to the next in the flow field. Also, it is important to note that the stagnation process alone is isentropic, and the flow need not be isentropic. Properties at the stagnation state are usually indicated with a subscript 0 *viz.,* P_0, T_0, ρ_0, and so

---

[2]Except, of course, at the point where the sonic state occurs.

on. Here P_0 is the stagnation pressure, T_0 is the stagnation temperature and ρ_0 is the stagnation density. Hereafter, P and T will be referred to as the static pressure and static temperature and the corresponding state point will be called the static state.

To derive the relationship between the static and stagnation states, we start by integrating Eq. 2.3 between these two states. This gives

$$\int_1^0 dh + \int_1^0 d\left(\frac{u^2}{2}\right) = 0$$

If we integrate this equation and rearrange, we get

$$h_0 = h_1 + \frac{u_1^2}{2} \tag{2.15}$$

after noting that the velocity is zero at the stagnation state. For a calorically perfect gas,[3] $dh = C_p dT$ and so

$$T_{0,1} - T_1 = \frac{u_1^2}{2C_p}$$

After using Eqs. 1.10 and 2.12, we can finally write

$$\frac{T_0}{T} = 1 + \frac{\gamma - 1}{2} M^2 \tag{2.16}$$

where the subscript for the static state has been dropped for convenience. Although the stagnation process is isentropic, this fact is not required for the calculation of stagnation temperature.

Since the stagnation process is isentropic, the static and stagnation states lie on the same isentrope. If we apply Eq. 2.12 between the static and stagnation states and use the fact that $s_0 = s_1$, we get

$$\frac{P_{0,1}}{P_1} = \left(\frac{T_{0,1}}{T_1}\right)^{\frac{\gamma}{\gamma - 1}}$$

If we substitute from Eq. 2.16, we get

---

[3]Equation 2.15 can be used even when the gas is not calorically perfect. This happens, for instance, when the temperatures encountered in a particular problem are outside the range in which the calorically perfect assumption is valid. In such cases, either the enthalpy of the gas is available as a function of temperature in tabular form or C_p is available in the form of a polynomial in temperature (see, for example, http://webbook.nist.gov). If the stagnation and static temperatures are known, then the velocity can be calculated from Eq. 2.15. On the other hand, if the static temperature and velocity are known, then the stagnation temperature has to be calculated either by tabular interpolation or iteratively starting with a suitable initial guess.

$$\frac{P_0}{P} = \left(1 + \frac{\gamma - 1}{2} M^2\right)^{\frac{\gamma}{\gamma - 1}} \tag{2.17}$$

where the subscript denoting the static state has been dropped. This equation can be derived in an alternative way, in a manner similar to the one used for the derivation of the stagnation temperature. This is somewhat longer but gives some interesting insights into the stagnation process. We start by rewriting Eq. 2.2 in the following form:

$$\frac{dP}{\rho} + d\left(\frac{u^2}{2}\right) = 0$$

By substituting Eq. 2.3, this can be simplified to read

$$\frac{dP}{\rho} - dh = 0$$

Integrating this between the static and stagnation states leads to

$$\int_1^0 \frac{dP}{\rho} - \int_1^0 dh = 0$$

Since the second term is a perfect differential, it can be integrated easily. The first term is not a perfect differential and so the integral depends on the path used for the integration—in other words, the path connecting states 1 and 0. Since this process is isentropic, from Eq. 2.6 we can show that

$$C_v \frac{dP}{P} + C_p \frac{dv}{v} = 0 \Rightarrow Pv^\gamma = constant = P_1 v_1^\gamma$$

Thus, the above equation reduces to

$$\int_1^0 \frac{P_1^{1/\gamma}}{\rho_1} \frac{dP}{P^{1/\gamma}} = C_p(T_{0,1} - T_1)$$

where we have invoked the calorically perfect gas assumption. With a little bit of algebra, this can be easily shown to lead to Eq. 2.17.

The stagnation density can be evaluated by using the equation of state $P_0 = \rho_0 R T_0$. Thus

$$\frac{\rho_0}{\rho} = \left(1 + \frac{\gamma - 1}{2} M^2\right)^{\frac{1}{\gamma - 1}} \tag{2.18}$$

Values of T_0/T, P_0/P and ρ_0/ρ for $\gamma = 1.4$ are listed in Table A.1 for M ranging from 0 to 5.

This derivation brings out the fact that unlike the stagnation temperature, the nature of the stagnation process has to be known in order to evaluate the stagnation pressure. This, in itself, arises from the fact that Eq. 2.2 is not a perfect differential. It would appear that we could have circumvented this difficulty by integrating Eq. 2.7 instead, which is a perfect differential. This would have led to the following expression:

$$P_{0,1} = P_1 + \rho_1 u_1^2$$

If we divide through by P_1 and use the fact that $P_1 = \rho_1 RT$ and $a_1 = \sqrt{\gamma RT_1}$, we get

$$\frac{P_0}{P} = 1 + \gamma M^2$$

This expression for stagnation pressure is disconcertingly (and erroneously!) quite different from Eq. 2.17. The inconsistency arises due to the use of the continuity equation while deriving Eq. 2.7. Continuity equation 2.1 is not applicable during the stagnation process, as otherwise $\rho_0 \to \infty$ as $u \to 0$. Hence, Eq. 2.7 is not applicable for the stagnation process.

Another important fact about stagnation quantities is that they depend on the frame of reference unlike static quantities which are frame independent. This is best illustrated through a numerical example.

Example 2.1 Consider the propagation of sound wave into quiescent air at 300 K and 100 kPa. With reference to Fig. 2.1, determine $T_{0,1}$ and $P_{0,1}$ in the stationary and moving frames of reference.

Solution In the stationary frame of reference, $u_1 = 0$ and so, $T_{0,1} = T_1 = 300$ K and $P_{0,1} = P_1 = 100$ kPa.

In the moving frame of reference, $u_1 = a_1$ and so $M_1 = 1$. Substituting this into Eqs. 2.16 and 2.17, we get $T_{0,1} = 360$ K and $P_{0,1} = 189$ kPa.

The difference between the values evaluated in different frames becomes more pronounced at higher Mach numbers.

As already mentioned, stagnation temperature and pressure are local quantities and so they can change from one point to another in the flow field.

Changes in stagnation temperature can be achieved by the addition or removal of heat or work.[4] Heat addition increases the stagnation temperature, while removal of heat results in a decrease in stagnation temperature. Changes in stagnation pressure

---

[4]In such cases, the energy equation has to be modified suitably. For example Eq. 2.3 will read as

$$dh + d\left(\frac{u^2}{2}\right) = \delta q - \delta w$$

and Eq. 2.10 will read as

$$h_1 + \frac{u_1^2}{2} = h_2 + \frac{u_2^2}{2} - Q + W$$

are brought about by work interaction or irreversibilities. Across a compressor where work is done on the flow, stagnation pressure increases while across a turbine where work is extracted from the fluid, stagnation pressure decreases. It is for this reason, that any loss of stagnation pressure in the flow is undesirable as it is tantamount to a loss of work. To see the effect of irreversibilities, we start with the last equality in Eq. 2.12 and substitute for T_2/T_1 and P_2/P_1 as follows:

$$\frac{T_2}{T_1} = \frac{T_2}{T_{0,2}} \frac{T_{0,2}}{T_{0,1}} \frac{T_{0,1}}{T_1}$$

and

$$\frac{P_2}{P_1} = \frac{P_2}{P_{0,2}} \frac{P_{0,2}}{P_{0,1}} \frac{P_{0,1}}{P_1}$$

From Eq. 2.12

$$s_2 - s_1 = R \ln \left[\left(\frac{T_2}{T_1}\right)^{\frac{\gamma}{\gamma - 1}} \Big/ \left(\frac{P_2}{P_1}\right) \right]$$

If we use Eqs. 2.16 and 2.17, we get

$$s_2 - s_1 = C_p \ln \frac{T_{0,2}}{T_{0,1}} - R \ln \frac{P_{0,2}}{P_{0,1}} \tag{2.19}$$

This equation shows that irreversibilities in an adiabatic flow lead to a loss of stagnation pressure, since, for such a flow, $s_2 > s_1$ and $T_{0,2} = T_{0,1}$ and so $P_{0,2} < P_{0,1}$. This equation also shows that heat addition in a compressible flow is always accompanied by a loss of stagnation pressure. Since, $T_{0,2} > T_{0,1}$ in this case, and $s_2 > s_1$, $P_{0,2}$ has to be less than $P_{0,1}$. These facts are important in the design of combustors and will be discussed later. This equation also shows that increase or decrease of stagnation pressure brought about through work interaction leads to a corresponding change in the stagnation temperature.

2.4   T-s and P-v Diagrams in Compressible Flows

T-s and P-v diagrams are familiar to most of the readers from their basic thermodynamics course. These diagrams are extremely useful in illustrating states and processes graphically. Both of these diagrams display the same information, since the thermodynamic state is fully fixed by the specification of two properties, either P, v

---

where q (and Q) and w (and W) refer to the heat and work interaction per unit mass. We have also used the customary sign convention from thermodynamics i.e., that heat added to a system is positive and work done by a system is positive.

Fig. 2.2 Constant pressure and constant volume lines on a T-s diagram; constant temperature and constant entropy lines on a P-v diagram

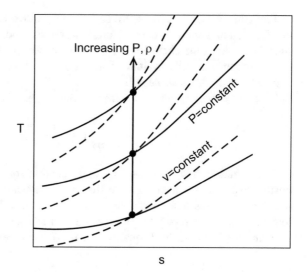

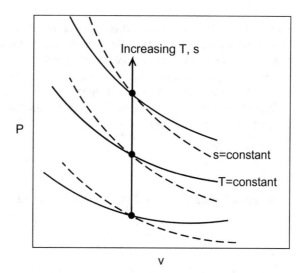

or T, s. Nevertheless, they are both useful as some processes can be depicted better in one than the other.

Let us review some basic concepts from thermodynamics in relation to T-s and P-v diagrams. Figure 2.2 shows thermodynamic states (filled circles) and contours of P, v (isobars and isochors) and contours of T, s (isotherms and isentropes). From the first equality in Eq. 2.6, we can write

$$dv = \frac{v}{R}ds - C_v\frac{v}{RT}dT \qquad (2.20)$$

From this equation, it is easy to see that, as we move along a $s = constant$ line in the direction of increasing temperature, v decreases, since, $dv = -(C_v/P)dT$, along such a line. Also, the change in v for a given change in T is higher at lower values of pressure than at higher values of pressure. This fact is of tremendous importance in compressible flows as we will see later.

Since $dv = 0$ along a $v = constant$ contour, from the above equation

$$\left.\frac{dT}{ds}\right|_v = \frac{T}{C_v} \tag{2.21}$$

This equation shows that the slope of the contours of v on a T-s diagram is always positive and is not a constant. Hence, the contours are not straight lines. Furthermore, the slope increases with increasing temperature and so the contours are shallow at low temperatures and become steeper at higher temperatures.

Similarly, from the third equality in Eq. 2.6, it can be shown that

$$\left.\frac{dT}{ds}\right|_P = \frac{T}{C_p} \tag{2.22}$$

for any isobar. The same observation made above regarding the slope of the contours of v on a T-s diagram are applicable to isobars as well. In addition, since $C_p > C_v$, at any state point, isochors are steeper than isobars on a T-s diagram and pressure increases along a $s = constant$ line in the direction of increasing temperature. These observations regarding isochors and isobars are shown in Fig. 2.2.

From the second equality in Eq. 2.6, the equation for isentropes on a P-v diagram can be obtained after setting $ds = 0$. Thus

$$\left.\frac{dP}{dv}\right|_s = -\frac{C_p}{C_v}\frac{P}{v} \tag{2.23}$$

By equating the second and the last term in Eq. 2.6, we get

$$C_v\frac{dT}{T} + R\frac{dv}{v} = C_p\frac{dT}{T} - R\frac{dP}{P}$$

This can be rearranged to give (after setting $dT = 0$)

$$\left.\frac{dP}{dv}\right|_T = -\frac{P}{v} \tag{2.24}$$

This equation shows that isotherms also have a negative slope on a P-v diagram and they are less steep than isentropes (Fig. 2.2). Furthermore, s and T increase with increasing pressure as we move along a $v = constant$ line.

Let us now look at using T-s and P-v diagrams for graphically illustrating states in one-dimensional compressible flows. In this case, in addition to T, s (or P, v),

Fig. 2.3 Illustration of states for a one-dimensional compressible flow on T-s and P-v diagram

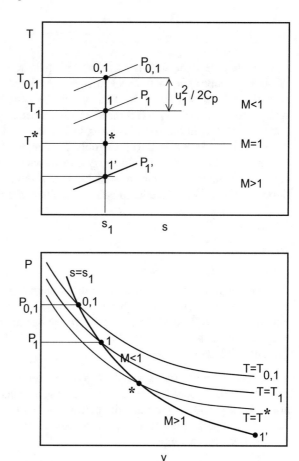

velocity information also has to be displayed. Equation 2.3 tells us how this can be done. For a calorically perfect gas, this can be written as

$$d\left(T + \frac{u^2}{2C_p}\right) = 0$$

Hence, at each state point, the static temperature is depicted as usual, and the quantity $u^2/2C_p$ is added to the ordinate (in case of a T-s diagram). Note that this quantity has units of temperature and the sum $T + u^2/2C_p$ is equal to the stagnation temperature T_0 corresponding to this state. This is shown in Fig. 2.3 for the subsonic state point marked 1. Also shown in this figure is the sonic state *corresponding to this state*. Once T_0 is known T^* can be evaluated from Eq. 2.16 by setting $M = 1$. Thus

$$\frac{T_0}{T^*} = \frac{\gamma + 1}{2}$$

Depicting the sonic state is useful since it tells at a glance whether the flow is subsonic or supersonic. All subsonic states will lie above the sonic state and all supersonic states will lie below. State point $1'$ shown in Fig. 2.3 is a supersonic state. This figure also shows that the stagnation process (1-0 or $1'$-0) is an isentropic process. All this information is shown in Fig. 2.3 on T-s as well as P-v diagram. Although it is conventional to show only T-s diagram in compressible flows, P-v diagrams are very useful when dealing with waves (for instance, shock waves and combustion waves). With this in mind, both the diagrams are presented side by side throughout, to allow the reader to become familiar with them.

Problems

2.1 Air enters the diffuser of an aircraft jet engine at a static pressure of 20 kPa and static temperature 217 K, and a Mach number of 0.9. The air leaves the diffuser with a velocity of 85 m/s. Assuming isentropic operation, determine the exit static temperature and pressure.

[249 K, 32 kPa]

2.2 Air is compressed adiabatically in a compressor from a static pressure of 100 kPa to 2000 kPa. If the static temperature of the air at the inlet and exit of the compressor are 300 K and 800 K, determine the power required per unit mass flow rate of air. Also, determine whether the compression process is isentropic or not.

[503 kW, Not isentropic]

2.3 Air enters a turbine at a static pressure of 2 MPa, 1400 K. It expands isentropically in the turbine to a pressure of 500 kPa. Determine the work developed by the turbine per unit mass flow rate of air and the static temperature at the exit.

[460 kW, 942 K]

2.4 Air at 100 kPa, 295 K and moving at 710 m/s is decelerated isentropically to 250 m/s. Determine the final static temperature and static pressure.

[515 K, 702 kPa]

2.5 Air enters a combustion chamber at 150 kPa, 300 K, and 75 m/s. Heat addition in the combustion chamber amounts to 900 kJ/kg. Air leaves the combustion chamber at 110 kPa and 1128 K. Determine the stagnation temperature, stagnation pressure, and velocity at the exit and the entropy change across the combustion chamber.

[1198 K, 136 kPa, 376 m/s, 1420 J/kg.K]

2.6 Air at 900 K and negligible velocity enters the nozzle of an aircraft jet engine. If the flow is sonic at the nozzle exit, determine the exit static temperature and velocity. Assume adiabatic operation.

[750 K, 549 m/s]

2.7 Air expands isentropically in a rocket nozzle from $P_0 = 3.5$ MPa, $T_0 = 2700$ K to an ambient pressure of 100 kPa. Determine the exit velocity, Mach number, and static temperature.

[1860 m/s, 2.97, 978 K]

2.8 Consider the capture stream tube of an aircraft engine cruising at Mach 0.8 at an altitude of 10 km. The capture mass flow rate is 250 kg/s. At station 1, which is in the freestream, the static pressure and temperature are 26.5 kPa and 223 K, respectively. At station 2, which is downstream of station 1, the cross-sectional area is 3 m². Further downstream at station 3, the Mach number is 0.4. Determine (a) the cross-sectional area at station 1 (usually called the capture area), (b) the Mach number at station 2, (c) the static pressure and temperature at stations 2 and 3, and (d) the cross-sectional at station 3.

[(a) 2.5213 m² (b) 0.5635 (c) Station2: 32.9 kPa, 236.52 K, Station3: 36.496 kPa, 243.74 K (d) 3.8258 m²]

2.9 The ramjet engine shown in Fig. 2.4 does not have any moving parts. It operates at high supersonic Mach numbers (< 4). The entering air is decelerated in the diffuser to a subsonic speed. Heat is added in the combustion chamber and the hot gases expand in the nozzle generating thrust. In an "ideal" ramjet engine, air is the working fluid throughout and the compression, expansion processes are isentropic. In addition, there is no loss of stagnation pressure due to heat addition. The air is expanded in the nozzle to the ambient pressure. Show that the Mach number of the air as it leaves the nozzle is the same as the Mach number of the air when it enters the diffuser. Sketch the process undergone by the air on T-s and P-v diagrams.

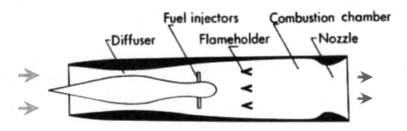

Fig. 2.4 Schematic of a ramjet engine

Chapter 3
Normal Shock Waves

Normal shock waves are compression waves that are seen in nozzles, turbomachinery blade passages, and shock tubes, to name a few. In the first two examples, normal shock usually occurs under off-design operating conditions or during start-up. The compression process across the shock wave is highly irreversible and so it is undesirable in such cases. In the last example, normal shock is designed to achieve extremely fast compression and heating of a gas with the aim of studying highly transient phenomena. Normal shocks are seen in external flows also. The term "normal" is used to denote the fact that the shock wave is normal (perpendicular) to the flow direction, before and after passage through the shock wave. This latter fact implies that there is no change in flow direction as a result of passing through the shock wave. In this chapter, we take a detailed look at the thermodynamic and flow aspects of normal shock waves.

3.1 Governing Equations

Figure 3.1 shows a normal shock wave propagating into quiescent air. The shock speed in the laboratory frame of reference is denoted as V_s. This figure is almost identical to Fig. 2.1, where the propagation of an acoustic wave is shown. The main differences are: (1) an acoustic wave travels with the speed of sound, whereas a normal shock travels at supersonic speeds and (2) the changes in properties across an acoustic wave are infinitesimal and isentropic, whereas they are large and irreversible across a normal shock wave.

If we switch to a reference frame in which the shock wave appears stationary, then the governing equations for the flow are Eqs. 2.8, 2.9, 2.10, and 2.12. These are reproduced here for convenience.

$$\rho_1 u_1 = \rho_2 u_2 \tag{2.8}$$

© The Author(s) 2021
V. Babu, *Fundamentals of Gas Dynamics*,
https://doi.org/10.1007/978-3-030-60819-4_2

Fig. 3.1 Illustration of a
normal shock wave

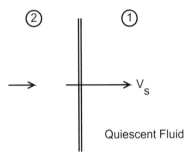

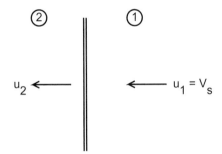

$$P_1 + \rho_1 u_1^2 = P_2 + \rho_1 u_2^2 \tag{2.9}$$

$$h_1 + \frac{u_1^2}{2} = h_2 + \frac{u_2^2}{2} \tag{2.10}$$

$$s_2 - s_1 = C_v \ln \frac{T_2}{T_1} + R \ln \frac{v_2}{v_1}$$

$$= C_v \ln \frac{P_2}{P_1} + C_p \ln \frac{v_2}{v_1} \tag{2.12}$$

$$= C_p \ln \frac{T_2}{T_1} - R \ln \frac{P_2}{P_1}$$

It can be seen from the energy equation that the stagnation temperature is constant across the shock wave, as there is no heat addition or removal.

3.2 Mathematical Derivation of the Normal Shock Solution

The continuity equation above can be written as

$$\frac{P_2}{P_1} = \sqrt{\frac{T_2}{T_1}} \frac{M_1}{M_2} \tag{3.1}$$

after using the fact that $u = M\sqrt{\gamma RT}$ and $\rho = P/RT$. Similarly, we can get from the momentum equation

$$\frac{P_2}{P_1} = \frac{1 + \gamma M_1^2}{1 + \gamma M_2^2} \tag{3.2}$$

and

$$\frac{T_2}{T_1} = \left(1 + \frac{\gamma - 1}{2} M_1^2\right) / \left(1 + \frac{\gamma - 1}{2} M_2^2\right) \tag{3.3}$$

from the energy equation. Combining these three equations, we get

$$\frac{1 + \frac{\gamma - 1}{2} M_1^2}{1 + \frac{\gamma - 1}{2} M_2^2} = \frac{M_2^2}{M_1^2} \left(\frac{1 + \gamma M_1^2}{1 + \gamma M_2^2}\right)^2$$

This is a quadratic equation in M_2^2. Given M_1, we can solve this equation to get M_2. With M_2 known, all the other properties at state 2 can be evaluated. This equation has only one meaningful solution, namely,

$$M_2^2 = \frac{2 + (\gamma - 1) M_1^2}{2\gamma M_1^2 - (\gamma - 1)} \tag{3.4}$$

The other solutions are either trivial ($M_2 = M_1$) or imaginary. Note that, if we set $M_1 = 1$ in Eq. 3.4 we get $M_2 = 1$, which is, of course, the solution corresponding to an acoustic wave. Also, a simple rearrangement of the expression in Eq. 3.4 shows that

$$M_2^2 = 1 - \frac{\gamma + 1}{2\gamma} \frac{M_1^2 - 1}{M_1^2 - 1 + \frac{\gamma + 1}{2\gamma}} \tag{3.5}$$

Hence, if $M_1 > 1$, then $M_2 < 1$ and vice versa. Thus, both the compressive solution $M_1 > 1$, $M_2 < 1$ and the expansion solution $M_1 < 1$, $M_2 > 1$ are allowed by the above equation. We must examine whether they are allowed based on entropy considerations. Since the process is adiabatic and irreversible, entropy has to increase across the shock wave. From Eq. 2.12, the entropy change across the shock wave is given as

$$s_2 - s_1 = C_p \ln \frac{T_2}{T_1} - R \ln \frac{P_2}{P_1}$$

Upon substituting the relations obtained above for T_2/T_1 and P_2/P_1, we get

$$s_2 - s_1 = C_p \ln \frac{M_2^2}{M_1^2} \left(\frac{1 + \gamma M_1^2}{1 + \gamma M_2^2} \right)^2 - R \ln \frac{1 + \gamma M_1^2}{1 + \gamma M_2^2}$$

This can be simplified to read

$$s_2 - s_1 = C_p \ln \frac{M_2^2}{M_1^2} + R \frac{\gamma + 1}{\gamma - 1} \ln \frac{1 + \gamma M_1^2}{1 + \gamma M_2^2}$$

Substituting for M_2 from Eq. 3.4, we get (after some tedious algebra!)

$$\frac{s_2 - s_1}{R} = \frac{1}{\gamma - 1} \ln \left[\frac{2\gamma M_1^2 - \gamma + 1}{\gamma + 1} \right] + \frac{\gamma}{\gamma - 1} \ln \left[\frac{2 + (\gamma - 1) M_1^2}{(\gamma + 1) M_1^2} \right]$$

With a slight rearrangement, this becomes

$$\frac{s_2 - s_1}{R} = \frac{1}{\gamma - 1} \ln \left[1 + \frac{2\gamma}{\gamma + 1} (M_1^2 - 1) \right]$$
$$+ \frac{\gamma}{\gamma - 1} \ln \left[1 - \frac{2}{\gamma + 1} \left(1 - \frac{1}{M_1^2} \right) \right]$$

It is clear from this expression that entropy across the shock wave increases when $M_1 > 1$ and decreases when $M_1 < 1$. Thus, for a normal shock, M_1 is always greater than one and M_2 is always less than one.

The static pressure and temperature can be seen to increase across the shock wave from Eqs. 3.2 and 3.3. Furthermore, from Eq. 3.1, it can be inferred that $P_2/P_1 > T_2/T_1$. It follows from this that

$$\frac{\rho_2}{\rho_1} = \left(\frac{P_2}{P_1} \right) \bigg/ \left(\frac{T_2}{T_1} \right) > 1 \tag{3.6}$$

Of course, due to the irreversibility associated with the shock, there is a loss of stagnation pressure. From Eq. 2.19, it is easy to show that

$$s_2 - s_1 = R \ln \frac{P_{0,1}}{P_{0,2}}$$

Thus, the stronger[1] the shock or higher the initial Mach number, the more the loss of stagnation pressure.

From Eq. 3.5, we get

$$M_2^2 = 1 - \frac{6}{7} \frac{M_1^2 - 1}{M_1^2 - \frac{1}{7}}$$

for diatomic gases for which $\gamma = 7/5$ and

$$M_2^2 = 1 - \frac{3}{5} \frac{M_1^2 - 1}{M_1^2 - \frac{2}{5}}$$

for monatomic gases for which $\gamma = 5/3$. A comparison of these two expressions suggests that for a given M_1, M_2 is higher for monatomic gases than diatomic gases. However, the strength of the shock as well as the temperature rise at a given M_1 is higher in the case of the former. This explains why monatomic gases are used extensively in shock tubes. Equations 3.4, 3.2, 3.3, 3.6 as well as the ratio $P_{0,2}/P_{0,1}$ are plotted in Fig. 3.2 for monatomic and diatomic gases.

In the limiting case when $M_1 = 1$, it is easy to see that the process is isentropic (as it should be, since it corresponds to the propagation of an acoustic wave). Also, $M_2 = 1$, $T_2/T_1 = 1$, $P_2/P_1 = 1$ and $\rho_2/\rho_1 = 1$ from Eqs. 3.5, 3.3, 3.2 and 3.6.

If we let $M_1 \to \infty$ in Eq. 3.5, then we have

$$M_2 = \sqrt{\frac{\gamma - 1}{2\gamma}}$$

$$\frac{P_2}{P_1} \to \infty$$

$$\frac{T_2}{T_1} \to \infty$$

$$\frac{\rho_2}{\rho_1} = \frac{\gamma + 1}{\gamma - 1}$$

These trends can be clearly seen in Fig. 3.2.

---

[1] Strength of a shock is usually defined as $\frac{P_2}{P_1} - 1$. .

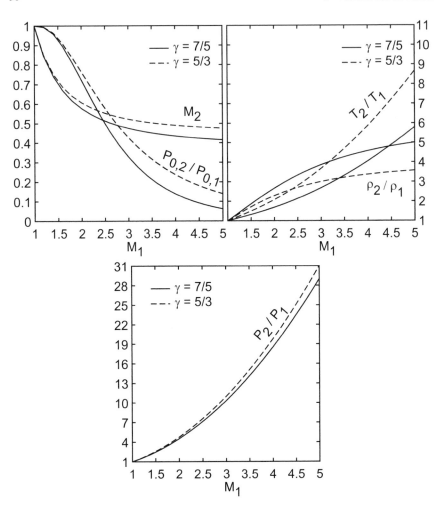

Fig. 3.2 Variation of the downstream Mach number and property ratios across a normal shock wave for monatomic and diatomic gases

3.3   Illustration of the Normal Shock Solution on T-s and P-v Diagrams

In this section, we will try to draw some insight into the normal shock compression process through graphical illustrations on the T-s and P-v diagrams. Figure 3.3 shows the T-s and P-v diagram for the normal shock process. The static (P_1, T_1), stagnation $(P_{0,1}, T_0)$, and sonic state (*) corresponding to state 1 are shown in this figure. State point 2 lies to the right of state point 1 (owing to the increase in entropy across the shock) and above the sonic state (since the flow becomes subsonic after the shock).

Fig. 3.3 Illustration of Normal Shock in T-s and P-v diagram

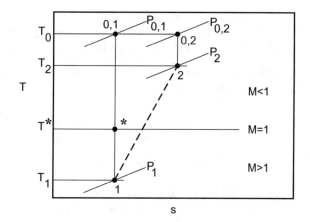

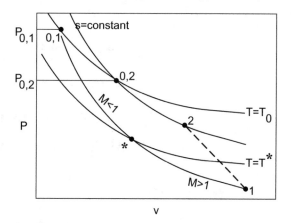

The corresponding stagnation state lies at the point of intersection of the isentrope (vertical line) through point 2 and the isotherm $T = T_0$. From the orientation of isobars in a T-s diagram (see Fig. 2.2), it is easy to see that the stagnation pressure corresponding to state point 2, $P_{0,2}$ is less than $P_{0,1}$. The normal shock itself is shown in this diagram as a heavy dashed line.

The same features are illustrated in a P-v diagram also in Fig. 3.3. Here, isotherms are shown as dashed lines and isentropes as solid lines. The stagnation state $(0, 1)$ lies at the point of intersection of the isentrope through state point 1 and the isotherm $T = T_0$. State point 2 lies to the left of and above point 1, since $v_2 < v_1$ and $P_2 > P_1$ and the increase in entropy causes it to lie on a higher isentrope. Since isentropes are steeper than isotherms, the isotherm $T = T_0$ intersects this isentrope at a lower value of pressure and so $P_{0,2} < P_{0,1}$.

Values of $M_2, P_2/P_1, T_2/T_1, \rho_2/\rho_1$, and $P_{0,2}/P_{0,1}$ are given in Table A.2 for values of M_1 from 1 to 5.

Example 3.1 Consider a normal shock wave that moves with a speed of 696 m/s into still air at 100 kPa and 300 K. Determine the static and stagnation properties ahead of and behind the shock wave in stationary and moving frames of reference.

Solution In a moving frame of reference in which the shock is stationary (observer moving with shock),

$$P_1 = 100\,\text{kPa}, \quad T_1 = 300\,\text{K}, \quad u_1 = 696\,\text{m/s}$$
$$a_1 = \sqrt{\gamma R T_1} = \sqrt{1.4 \times 288 \times 300} = 348\,\text{m/s}$$
$$M_1 = 2, \quad T_{0,1} = 540\,\text{K}, \quad P_{0,1} = 782.4\,\text{kPa}$$

We can use Eq. 3.4 to evaluate M_2 or use Table A.2. The latter choice allows us to look up pressure ratio, temperature ratio, and other ratios, in one go. For $M_1 = 2$, from normal shock table, we get

$$\frac{P_2}{P_1} = 4.5 \Rightarrow P_2 = 450\,\text{kPa}$$

$$\frac{T_2}{T_1} = 1.687 \Rightarrow T_2 = 506\,\text{K}$$

$$\frac{P_{0,2}}{P_{0,1}} = 0.7209 \Rightarrow P_{0,2} = 564\,\text{kPa}$$

$$M_2 = 0.5774 \Rightarrow u_2 = 260.8\,\text{m/s}$$

Switching now to a stationary frame of reference (observer stationary) in which the shock moves with speed $V_s = 696 m/s$,

$$P_1 = 100\,\text{kPa}, \quad T_1 = 300\,\text{K}, \quad u_1 = 0\,\text{m/s}$$
$$P_{0,1} = 100\,\text{kPa}, \quad T_{0,1} = 300\,\text{K}$$
$$P_2 = 450\,\text{kPa}, \quad T_2 = 506\,\text{K}$$
$$u_2 = 696 - 260.8 = 435.2\,\text{m/s}$$

$$M_2 = 435.2/\sqrt{1.4 \times 288 \times 506} = 0.9635$$
$$T_{0,2} = 600\,\text{K}, \quad P_{0,2} = 817\,\text{kPa}$$

These numbers are shown in Fig. 3.4, to illustrate them more clearly. Note that, in the moving frame of reference, stagnation temperature remains constant while stagnation pressure decreases. On the other hand, in the stationary frame, *both* stagnation temperature and stagnation pressure increase. This clearly shows the frame dependence of the stagnation quantities.

Observer Stationary

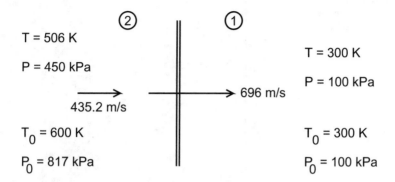

Observer Moving

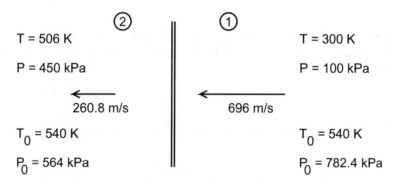

Fig. 3.4 Worked example showing static and stagnation properties in stationary and moving frames of reference

3.4 Further Insights into the Normal Shock Wave Solution

In this section, further insights into the normal shock wave solution are presented. The methodology is quite useful in the study of not only normal shock waves, but combustion waves also.

We start by writing the continuity equation Eq. 2.8 as follows:

$$\rho_1 u_1 = \rho_2 u_2 = \dot{m}/A = G \tag{3.7}$$

Fig. 3.5 Illustration of
Rayleigh line and H-curve.
For a given initial state 1,
final states in the shaded
regions are forbidden.

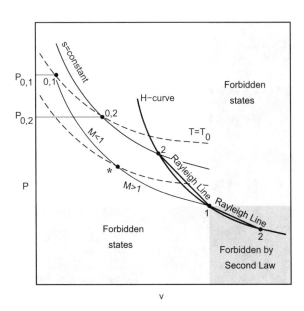

where $\dot{m}$ is the mass flow rate, A is the cross-sectional area and $G(> 0)$ is a constant.
Substituting for u_1 and u_2 from this equation into the momentum equation Eq. 2.9,
we get

$$P_1 + G^2 v_1 = P_2 + G^2 v_2$$

where we have used the fact that $\rho = 1/v$. This can be rewritten as

$$\frac{P_1 - P_2}{v_1 - v_2} = -G^2 \tag{3.8}$$

This is the equation for a straight line with slope $-G^2$ in P-v coordinates. This line
is referred to as the Rayleigh line. This is shown as a thick line in Fig. 3.5. Note that
since $G > 0$, the slope of the Rayleigh line is always negative and so downstream
states are constrained to lie in the second and fourth quadrants with respect to the
initial state 1. Hence, state 2 cannot lie in the shaded regions in Fig. 3.5. Since G
is a real quantity, Eq. 3.8 also allows a compressive solution ($P_2 > P_1$ and $v_2 < v_1$)
which lies in the second quadrant and an expansion solution ($P_2 < P_1$ and $v_2 > v_1$)
which lies in the fourth quadrant. As we already showed in the previous section,
only the former solution is allowed by second law of thermodynamics. Thus, state 2
cannot lie in the fourth quadrant also in Fig. 3.5.

If we rewrite the energy equation, Eq. 2.10 in the same manner in terms of P and
v, we get

$$\frac{\gamma R}{\gamma - 1}T_1 + \frac{1}{2}v_1^2 G^2 = \frac{\gamma R}{\gamma - 1}T_2 + \frac{1}{2}v_2^2 G^2 \tag{3.9}$$

Upon rearranging, we get

$$\frac{\gamma R}{\gamma - 1}T_1\left(1 - \frac{T_2}{T_1}\right) = -\frac{1}{2}(v_1^2 - v_2^2)G^2$$

If we substitute for $-G^2$ from Eq. 3.8, we get

$$\frac{\gamma R}{\gamma - 1}T_1\left(1 - \frac{T_2}{T_1}\right) = \frac{1}{2}(v_1 + v_2)(v_1 - v_2)\frac{P_1 - P_2}{v_1 - v_2}$$

Simplifying

$$\frac{\gamma R}{\gamma - 1}T_1\left(1 - \frac{T_2}{T_1}\right) = \frac{1}{2}v_1\left(1 + \frac{v_2}{v_1}\right)P_1\left(1 - \frac{P_2}{P_1}\right)$$

From the equation of state, $P_1 v_1 = RT_1$ and $T_2/T_1 = P_2 v_2 / P_1 v_1$. Thus,

$$\frac{\gamma}{\gamma - 1}\left(1 - \frac{P_2}{P_1}\frac{v_2}{v_1}\right) = \frac{1}{2}\left(1 + \frac{v_2}{v_1}\right)\left(1 - \frac{P_2}{P_1}\right)$$

By rearranging and grouping terms, it is easy to show that

$$\frac{P_2}{P_1} = \left(\frac{v_2}{v_1} - \frac{\gamma + 1}{\gamma - 1}\right) \Big/ \left(1 - \frac{v_2}{v_1}\frac{\gamma + 1}{\gamma - 1}\right) \tag{3.10}$$

This equation is called the Rankine–Hugoniot equation. It is the equation for a quadratic in P-v coordinates, called the H-curve and is shown in Fig. 3.5. The points of intersection (state points 1 and 2) of the Rayleigh line and the H-curve in the P-v diagram represent the normal shock solution.

3.4.1 Normal Shock Compression and Isentropic Compression

It is not difficult to see that the H-curve (Eq. 3.10) is steeper than the isentrope that passes through state 1. So, for a given change in specific volume, the normal shock process can achieve a higher compression than an isentropic process, albeit with a loss of stagnation pressure.

It can be seen from Fig. 3.6 that for given values of P_1, v_1, and v_2, normal shock compression results in a higher value for P_2 than isentropic compression (in which the pressure increases to P_{2s}). It is also evident from this figure that, for given values of P_1, v_1 and P_2, the specific volume at the end of the isentropic compression

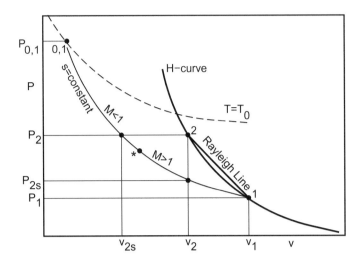

Fig. 3.6 Illustration of normal shock compression and isentropic compression

process, v_{2s}, is less than that at the end of normal shock compression, v_2. Both these situations, namely, compression between given specific volumes or a given pressure rise, are relevant in practical applications. It is clear that normal shock compression is more effective but less efficient than isentropic compression. The former attribute is of importance in intakes of supersonic vehicles, since it determines the length of the intake. However, the latter attribute is also important as loss of stagnation pressure is tantamount to loss of work and so an optimal operating condition has to be determined. An inspection of Fig. 3.2 reveals that the loss of stagnation pressure is about 20% for $M_1 = 2$ and about 70% for $M_1 = 3$. This suggests that compression using normal shocks is both effective and reasonably efficient for $M_1 \leq 2$. Accordingly, in supersonic intakes, the flow is decelerated to this value using other means and the compression process is terminated using a normal shock.

The H-curve passing through state point 1, is the locus of all possible downstream states, some allowed and others not allowed. The actual downstream state for a given value of G, is fixed by the Rayleigh line passing through state point 1. For some values of G, the Rayleigh line drawn from point 1 may not intersect the H-curve at all,[2] which shows that a normal shock solution is not possible for such cases.

We close this chapter with a very insightful illustration of the situation shown at the top in Figs. 2.1 and 3.1.[3] The ambient air at a certain location which is initially at the quiescent condition in both these situations is shown in Fig. 3.7 as state *1*. After the passage of the normal shock wave, the air at this location is almost instantaneously taken to state *2*, since the thickness of the shock wave is negligibly small. The air which is now at a higher pressure, higher temperature, and higher density relaxes to

---

[2]except trivially at point 1 itself.

[3]"Shock waves versus Sound waves", *LOS ALAMOS SCIENCE*, Spring/Summer 1985, pp. 42–43.

Fig. 3.7 Illustration of the effect of passage of a normal shock wave and a sound wave

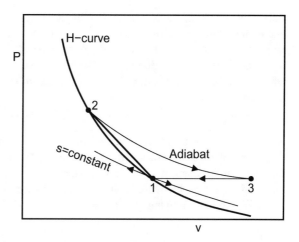

the ambient pressure adiabatically (state *3*), on account of the fact that this expansion process from such a high pressure is extremely rapid and hence there is very little time for heat exchange with the surrounding ambient to take place. Thermal relaxation back to the initial state *1* follows much more slowly.

On the other hand, as a result of passage of a sound wave, the state of the air at the given location oscillates back and forth about state *1* along the isentrope that passes through it. The frequency of this oscillation is, of course, the same as the frequency of the sound wave. The extent of the excursion from the mean state (*1*), is infinitesimally small, as discussed in the previous chapter. The arrows shown in Fig. 3.7 along the isentrope passing through state *1* are highly exaggerated for the sake of clarity. Unlike a normal shock wave, which is just a discontinuity, a sound wave consists of alternating compression and rarefaction fronts. Passage of a compression front causes the state of the fluid to move from *1* along the isentrope to the left and above and then relax back to *1*. Passage of the subsequent rarefaction front causes movement of the state to the right and below along the isentrope and back.

Problems

3.1 A shock wave advances into stagnant air at a pressure of 100 kPa and 300 K. If the static pressure downstream of the wave is tripled, what are the shock speed and the absolute velocity of the air downstream of the shock?

[573 m/s, 302 m/s]

3.2 Repeat Problem 1 assuming the fluid to be helium instead of air.

[1644.97 m/s, 757.49 m/s.]

3.3 Air at 2.5 kPa, 221 K approaches the intake of a ramjet engine operating at an altitude of 25 km. The Mach number is 3.0. For this Mach number, a normal shock stands just ahead of the intake. Determine the stagnation pressure, static pressure, and temperature of the air immediately after the normal shock. Also, calculate the percent loss in stagnation pressure. Repeat the calculations for Mach number equal to 4. The high loss of stagnation pressure that you see from your calculations illustrates why the intake of a ramjet has to be designed carefully to avoid such normal shocks during operation.

[30 kPa, 26 kPa, 592 K, 67%; 53 kPa, 46 kPa, 894 K, 86%]

3.4 A blast wave passes through still air at 300 K. The velocity of the air behind the wave is measured to be 180 m/s in the laboratory frame of reference. Determine the speed of the blast wave in the laboratory frame of reference and the stagnation temperature behind the wave in the laboratory as well moving frames of reference. You will find the following relations useful:

$$\frac{P_2}{P_1} = \frac{\gamma - 1}{\gamma + 1} \left[\frac{2\gamma M_1^2}{\gamma - 1} - 1 \right]$$

$$\frac{T_2}{T_1} = \left(1 + \frac{\gamma - 1}{2} M_1^2\right) \left(\frac{2\gamma M_1^2}{\gamma - 1} - 1\right) \left[M_1^2 \frac{(\gamma + 1)^2}{2(\gamma - 1)}\right]^{-1}$$

[472.6 m/s, 385 K, 411 K]

3.5 A normal shock wave travels into still air at 300 K. If the static temperature of the air is increased by 50 K as a result of the passage of the shock wave, determine the speed of the wave in the laboratory frame of reference.

[437.46 m/s]

3.6 A shock wave generated due to an explosion travels at a speed of 1.5 km/s into still air at 100 kPa and 300 K. Determine the velocity of the air, static, and stagnation quantities (with respect to a stationary frame of reference) in the region through which the shock has passed.

[1183 m/s, 2.2 MPa, 1370 K, 9.1 MPa, 2067 K]

3.7 A bullet travels through air (300 K, 100 kPa) at twice the speed of sound. Determine the temperature and pressure at the nose of the bullet.

Note that although there will be a curved, bow shock ahead of the bullet, in the nose region, normal shock relationships can be used. Also note that the nose is a stagnation point!

[540 K, 565 kPa]

3.8 A pitot tube is used to measure the Mach number (M_1) of a supersonic flow as shown in the figure.

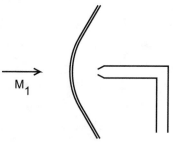

Although a curved shock stands ahead of the probe, it is fairly accurate to assume that the fluid in the stream tube captured by the probe has passed through a normal shock wave. It is also reasonable to assume that the probe measures the stagnation pressure downstream of the shock wave ($P_{0,2}$). If the static pressure upstream of the shock wave (P_1) is also measured, then the Mach number M_1 can be evaluated. Derive the relation connecting $P_{0,2}/P_1$ and M_1 (this is called the Rayleigh pitot formula).

Chapter 4
Flow with Heat Addition—Rayleigh Flow

In this chapter, we look at one-dimensional flow in a constant area duct with heat addition. Heat interaction would be more appropriate, since the theory that is developed applies equally well to situations where heat is removed. However, such a situation is rarely, if ever, encountered. Hence the predominant interest is on flows with heat addition, which are encountered in combustors ranging from those in aviation gas turbine engines through ramjet engines to scramjet engines. The corresponding combustor entry Mach number in these applications range from low subsonic through high subsonic to supersonic.

4.1 Governing Equations

The governing equations for this flow are Eqs. 2.8, 2.9, and 2.12,

$$\rho_1 u_1 = \rho_2 u_2 \tag{2.8}$$

$$P_1 + \rho_1 u_1^2 = P_2 + \rho_2 u_2^2 \tag{2.9}$$

$$s_2 - s_1 = C_v \ln \frac{P_2}{P_1} + C_p \ln \frac{v_2}{v_1} \tag{2.12}$$

The energy equation Eq. 2.10 has to be modified slightly to account for heat interaction and so

$$h_1 + \frac{u_1^2}{2} + q = h_2 + \frac{u_2^2}{2} \tag{4.1}$$

© The Author(s) 2021
V. Babu, *Fundamentals of Gas Dynamics*,
https://doi.org/10.1007/978-3-030-60819-4_4

where q is the heat interaction per unit mass and is positive when heat is added to the flow and negative when heat is removed. Upon using the calorically perfect gas assumption and the definition of the stagnation temperature, $T_0 = T + u^2/2C_p$, we get

$$T_{0,2} - T_{0,1} = \frac{q}{C_p} \tag{4.2}$$

This equation shows that addition of heat to a flow increases the stagnation temperature, while heat removal decreases it.

Although the governing equations for this flow resemble those presented in the previous chapter for normal shock waves, the important difference lies in the solution. The flow properties now change uniformly along the length of the duct, whereas in the former case, there is a discontinuity across the shock wave. It is also of interest to note that the spatial coordinate does not appear anywhere in the above equations. Thus, state points 1 and 2 represent the conditions at the entrance and the exit of the duct.

4.2 Illustration on T-s and P-v Diagrams

Before we discuss the solution procedure for solving the above equations, let us try to get some physical intuition on the solution to Eqs. 2.8, 2.9, and 2.12. Starting from the inlet state, we will take a small step corresponding to the addition or removal of an incremental amount of heat δq and try to determine the next state as dictated by these equations. Successive steps will then allow us to determine the locus of all the allowed downstream states. To this end, we will relate changes in all the properties to δq and determine the next state point on the T-s diagram. Since the addition or removal heat results in a corresponding change in the stagnation temperature, it is more convenient to relate the change in properties to the change in the stagnation temperature dT_0.

From Eq. 2.1, we get

$$\frac{d\rho}{\rho} = -\frac{du}{u} \tag{4.3}$$

From Eq. 2.7, we get

$$dP = -\rho u du$$

Since $P = \rho RT$ and $a^2 = \gamma RT$, this can be written as

$$\frac{dP}{P} = -\gamma M^2 \frac{du}{u} \tag{4.4}$$

From the equation of state $P = \rho RT$, we get

$$dT = \frac{1}{\rho R} dP - \frac{T}{\rho} d\rho$$

Substituting for dP and $d\rho$ from above and simplifying, we get

$$\frac{dT}{T} = (1 - \gamma M^2) \frac{du}{u} \tag{4.5}$$

Equation 2.6 can be written as

$$ds = C_v \frac{dP}{P} - C_p \frac{d\rho}{\rho}$$

If we substitute for dP and $d\rho$ from above, we get

$$ds = C_v \gamma (1 - M^2) \frac{du}{u} \tag{4.6}$$

From the definition of stagnation temperature,

$$dT_0 = dT + \frac{1}{C_p} u du$$

This can be simplified to read as

$$dT_0 = (1 - M^2) T \frac{du}{u}$$

first and then as

$$\frac{dT_0}{T_0} = \frac{1 - M^2}{1 + \frac{\gamma - 1}{2} M^2} \frac{du}{u} \tag{4.7}$$

Finally, from the definition of Mach number, $M = u/\sqrt{\gamma RT}$, we can write

$$dM = M \frac{du}{u} - \frac{M}{2} \frac{dT}{T}$$

This can be simplified to read

$$\frac{dM}{M} = \frac{1 + \gamma M^2}{2} \frac{du}{u} \tag{4.8}$$

If we use Eq. 4.7 to eliminate du/u in favor of dT_0/T_0 in Eqs. 4.3–4.6 and 4.8, we finally get

Table 4.1 Changes in properties for a given change in T_0

	$T_0 \uparrow (q > 0)$	$T_0 \downarrow (q < 0)$
$M < 1$	$u \uparrow \rho \downarrow P \downarrow T \uparrow s \uparrow M \uparrow$	$u \downarrow \rho \uparrow P \uparrow T \downarrow s \downarrow M \downarrow$
$M > 1$	$u \downarrow \rho \uparrow P \uparrow T \uparrow s \uparrow M \downarrow$	$u \uparrow \rho \downarrow P \downarrow T \downarrow s \downarrow M \uparrow$

$$\frac{d\rho}{\rho} = -\frac{1 + \frac{\gamma-1}{2}M^2}{1 - M^2}\frac{dT_0}{T_0}$$

$$\frac{dP}{P} = -\gamma M^2 \frac{1 + \frac{\gamma-1}{2}M^2}{1 - M^2}\frac{dT_0}{T_0}$$

$$\frac{dT}{T} = (1 - \gamma M^2)\frac{1 + \frac{\gamma-1}{2}M^2}{1 - M^2}\frac{dT_0}{T_0} \tag{4.9}$$

$$ds = C_v \gamma \left(1 + \frac{\gamma - 1}{2}M^2\right)\frac{dT_0}{T_0}$$

$$\frac{du}{u} = \frac{1 + \frac{\gamma-1}{2}M^2}{1 - M^2}\frac{dT_0}{T_0}$$

$$\frac{dM}{M} = \frac{1 + \gamma M^2}{2}\frac{1 + \frac{\gamma-1}{2}M^2}{1 - M^2}\frac{dT_0}{T_0}$$

Let us now tabulate the changes in properties from Eq. 4.9 for a given change in dT_0. These changes are shown symbolically in Table 4.1, using $\uparrow$ and $\downarrow$ to indicate increasing and decreasing trends.

The observations in Table 4.1 can be summarized conveniently in terms of heat addition (increasing T_0) and heat removal (decreasing T_0) as follows. When heat is added to a subsonic flow, static temperature, velocity, Mach number, and entropy increase. Thus the next state point lies to the right and above at a lower static pressure and density on a T-s diagram. On the other hand, when heat is added to a supersonic flow, static temperature, and entropy increase, while velocity and Mach number decrease. Thus the next state point lies to the right and above as before but at a higher static pressure and density on a T-s diagram. In both cases, heat removal shows the exact opposite trend.

Furthermore, upon combining Eqs. 4.5 and 4.6, we get

$$\frac{dT}{ds} = \frac{1 - \gamma M^2}{\gamma(1 - M^2)}\frac{T}{C_v} \tag{4.10}$$

which is the slope of the Rayleigh curve. The following inferences can be drawn from Eq. 4.10.

- The slope of the Rayleigh curve is positive for supersonic Mach numbers *i.e.*, $dT/ds > 0$ when $M > 1$.
- The supersonic branch of the Rayleigh curve at any point is steeper than the isobar and the isochor passing through the same point (Eqs. 2.21 and 2.22).
- The slope $dT/ds \to +\infty$, as $M \to 1$ from an initially supersonic Mach number.
- The slope of the Rayleigh curve is positive for subsonic Mach numbers in the range $0 < M \leq 1/\sqrt{\gamma}$ and negative in the range $1/\sqrt{\gamma} < M \leq 1$.
- The subsonic branch of the Rayleigh curve at any point is less steep than the isobar and the isochor passing through the same point (Eqs. 2.21 and 2.22).
- The slope $dT/ds \to -\infty$, as $M \to 1$ from an initially subsonic Mach number.
- When $1/\sqrt{\gamma} < M < 1$, the static temperature actually *decreases* with heat addition.
- Entropy reaches a maximum at the sonic state.

These findings allow us to construct the locus of all the possible states (for the given inlet state or mass flow rate) and ultimately the state at the end of the heat interaction process, step by step. This curve is called the Rayleigh curve and is illustrated in Fig. 4.1. This is the same as the Rayleigh line encountered before, but in the T-s plane instead of the P-v plane.

The sonic state represents a limiting state for both subsonic and supersonic initial states. The amount of heat necessary to go from a given initial state to the sonic state represents the maximum amount of heat that can be added from this initial state. Of significance is the fact that such a limitation is present only for heat addition, not heat removal.

In principle, starting from a state in the subsonic or supersonic portion of the Rayleigh curve, it is possible to traverse through the sonic state onto the other branch with the appropriate combination of heat interaction. Such an arrangement is not practical, however.

In Sect. 1.1, it was mentioned that, in the absence of compressibility effects, heat addition results in a change in temperature and a change in density arising from it, without a change in pressure. This can be clearly demonstrated from Eq. 4.9 in the incompressible limit *i.e.*, by letting $M \to 0$. This leads to

$$\frac{d\rho}{\rho} = -\frac{dT_0}{T_0}; \quad \frac{dP}{P} = 0; \quad \frac{dT}{T} = \frac{dT_0}{T_0}$$

When compressibility effects are present, Eq. 4.9 shows that heat addition results in a change in temperature, density as well as pressure.

An important fact with respect to heat addition is that it always results in a loss of stagnation pressure. If we assume that the heat addition is a reversible process, then from Eq. 2.4, we get

$$ds = \frac{C_p dT_0}{T}$$

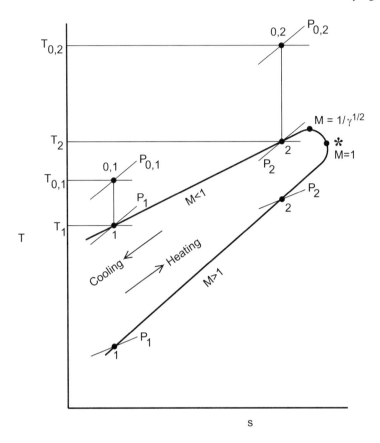

Fig. 4.1 Illustration of Heat Interaction on T-s diagram

where we have used the fact that $\delta q = C_p dT_0$ from Eq. 4.2. Using Eq. 2.16, the right-hand side can be rewritten as

$$ds = C_p \left(1 + \frac{\gamma - 1}{2} M^2 \right) \frac{dT_0}{T_0}$$

From Eq. 2.19, we can write

$$ds = C_p \ln \frac{T_0 + dT_0}{T_0} - R \ln \frac{P_0 + dP_0}{P_0}$$

which, for $dT_0 \ll 1$ and $dP_0 \ll 1$, reduces to

$$ds = C_p \frac{dT_0}{T_0} - R \frac{dP_0}{P_0}$$

If we eliminate ds from these expressions, we get

$$\frac{dP_0}{P_0} = -\frac{\gamma M^2}{2}\frac{dT_0}{T_0}$$

We can also show this in an alternate way by starting with the relationship

$$\frac{P_0}{P} = \left(\frac{T_0}{T}\right)^{\frac{\gamma}{\gamma-1}}$$

If we take the logarithm on both sides and rearrange, we get

$$\ln P_0 = \ln P + \frac{\gamma}{\gamma-1}\ln T_0 - \frac{\gamma}{\gamma-1}\ln T$$

Upon taking the differential of this equation, and substituting for the differentials in the right-hand side from Eq. 4.9, we can show after a little bit of algebra that

$$\frac{dP_0}{P_0} = -\frac{\gamma M^2}{2}\frac{dT_0}{T_0} \tag{4.11}$$

From this expression, it is clear that whenever dT_0 is positive, dP_0 is negative for both subsonic and supersonic flow. This loss of stagnation pressure is inherent to heat addition and does not arise due to any irreversibility (it may be recalled that we assumed at the beginning that the heat addition process is reversible). Hence, this is an important factor in the design of combustors, whether subsonic or supersonic. Good mixing of the fuel and air is essential for good combustion, but this only contributes further to the loss of stagnation pressure since mixing is a highly irreversible process. Such conflicting factors make the design of combustors a very challenging task.

Let us now turn to the illustration of the heat interaction process on a P-v diagram. The equation of the Rayleigh line is the same as before, as there is no change in the momentum equation. The H-curve derived before corresponds to the special case $q = 0$, whereas, now q can be nonzero. This, however, does not alter the nature of the curve and we now have a family of H-curves, one for each value of q, as shown in Fig. 4.2. Starting from the energy equation 4.1 and using the same algebra as used in the previous chapter, it is easy to show that, the H-curve is given as

$$\frac{P_2}{P_1} = \left(\frac{v_2}{v_1} - \frac{\gamma+1}{\gamma-1} - \frac{2q}{RT_1}\right) \Big/ \left(1 - \frac{v_2}{v_1}\frac{\gamma+1}{\gamma-1}\right) \tag{4.12}$$

Downstream states that lie in the fourth quadrant are allowed now, consistent with the required changes in entropy as a result of the heat interaction. Four Rayleigh lines are shown in this figure, two corresponding to $M_1 > 1$ (thick solid lines) and two corresponding to $M_1 < 1$ (thick broken lines). The observations made above regarding the properties of the downstream state can be seen here as well, when one keeps in mind the nature of isotherms and isentropes on the P-v diagram (Fig. 3.4).

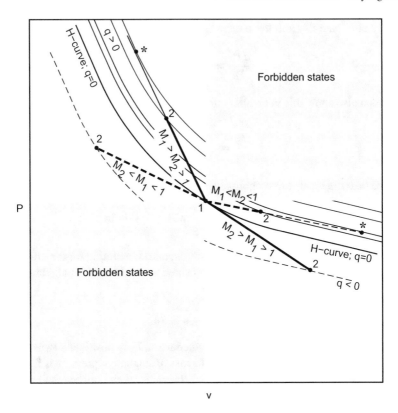

Fig. 4.2 Illustration of Heat Interaction on P-v diagram

In the case of heat addition, the Rayleigh line drawn from point 1 becomes tangential to the H-curve corresponding to a particular value of q, say q^*. Since the downstream state must lie at the point of intersection of an H-curve and the Rayleigh line corresponding to the given value of G, it is clear from this geometric construction that q^* represents a limiting value. State point 2 then would lie at the point of tangency. In other words, the slope of the H-curve (dP/dv) at this point is equal to $-G^2$, the slope of the Rayleigh line. We can show that this point is the sonic state point, starting from Eq. 3.9 and using the equation of state:

$$\frac{\gamma}{\gamma - 1} P_1 v_1 + \frac{1}{2} v_1^2 G^2 + q^* = \frac{\gamma}{\gamma - 1} P v + \frac{1}{2} v^2 G^2$$

where we have allowed the second state to be anywhere along the H-curve. Differentiating this equation with respect to v, we get

$$\frac{\gamma}{\gamma - 1} \left(v \frac{dP}{dv} + P \right) + v G^2 = 0$$

Thus

$$\frac{dP}{dv} = -\frac{\gamma - 1}{\gamma} G^2 - \frac{P}{v}$$

Equating this to $-G^2$, the slope of the Rayleigh line, leads to

$$-\frac{\gamma - 1}{\gamma} G^2 - \frac{P}{v} = -G^2$$

Or

$$\frac{P}{v} = \frac{G^2}{\gamma}.$$

Substituting $G = \rho u$ and $v = 1/\rho$, we are finally led to the result that at the point of tangency,

$$u_2 = \sqrt{\frac{\gamma P}{\rho}} = a_2$$

Hence, for this limiting value of G, the downstream Mach number M_2 is unity. This can be generalized to state that at the point of tangency of the Rayleigh line and the H-curve, the Mach number is always equal to one.

4.3 Thermal Choking and Its Consequences

As we have already seen, if, for a given mass flow rate, the heat added is equal to q^*, then the Mach number at the duct exit becomes 1. The duct is then said to be *choked*. Choking can happen in a flow due to several reasons. Since, in this case, the choking is due to heat addition, it is called *thermal choking*. Once the flow is thermally choked, further heat addition is not possible. The question that arises then is, what would happen if further heat were added? The exact answer to this question depends upon whether the flow is subsonic or supersonic at the inlet. However, there are drastic changes in the flow field due to further heat addition.

When the heat added is more than the q^* for the given inlet conditions, the point of intersection of the Rayleigh line with the H-curve for this value of heat addition lies beyond the point of tangency on the H-curve corresponding to q^*. In a flow with continuous heat addition such as the present one, states on the Rayleigh line beyond the sonic state are not accessible. Figure 4.3 shows that a Rayleigh line actually intersects an H-curve at two points. The first point alone is accessible through continuous heat addition, for the reason mentioned above. However, in the case of a discontinuity such as a combustion wave, the other state point is also accessible.

Fig. 4.3 Points of
intersection between a
Rayleigh line and a H-curve

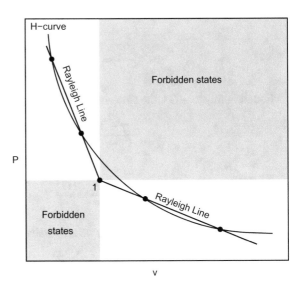

Fig. 4.4 a Illustration of
heat addition process with
$q > q^*$ for $M_1 < 1$.

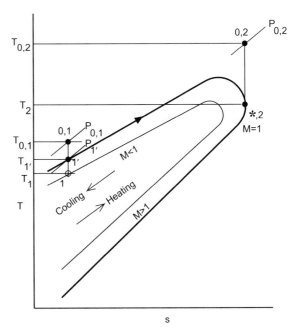

Here, we can directly jump to the other state point without going through any of the
intermediate points similar to a normal shock wave.

One way out of this situation is to operate along a Rayleigh line with a lesser
slope. Since the slope of the Rayleigh line is $-G^2 = -(\dot{m}/A)^2$, this means that the
mass flow rate has to be reduced keeping duct area the same or the duct area has to

Fig. 4.4 b Illustration of heat addition process with $q > q^*$ for $M_1 > 1$.

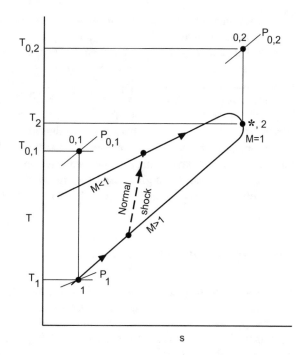

be increased keeping the mass flow rate the same. The former, of course, requires the inlet static conditions to be different and so we have to shift to a different Rayleigh curve (in the T-s diagram) or Rayleigh line (in the P-v diagram). Such a change in inlet static conditions is possible only if the flow is subsonic. This is illustrated in Fig. 4.4a, where the inlet state moves from state 1 (open circle) to 1′ (filled circle), to accommodate the heat release. On the other hand, if the flow is supersonic at the inlet, the static pressure increases due to the heat addition, and so a normal shock stands somewhere along the duct. Since the flow is subsonic after the shock, the state point moves to the subsonic portion of the (same) Rayleigh curve. The Mach number continues to increase due to the heat addition. The exact location of the shock wave depends upon the inlet Mach number and the amount of heat added. However, the flow in such a situation is no longer steady and the normal shock keeps moving upstream and eventually positions itself outside the duct. This is not tantamount to a change in inlet conditions as in the subsonic case; it means that the required pressure rise can be achieved only in this manner. The process corresponding to the supersonic inlet cases is illustrated in Fig. 4.4b. Although the exit state is shown to be the sonic state in Fig. 4.4a, b, the actual exit state will be such that the exit static pressure matches the ambient pressure in the region into which the duct exhausts and the exit Mach number will be less than unity. However, if the value of the ambient pressure is too low, then, the exit Mach number will be equal to unity with the exit static pressure being more than the ambient value.[1]

In real applications such as aircraft gas turbine engines, excess heat addition and the consequent adjustment in mass flow rate can result in highly undesirable pressure oscillations. In the case of ramjets and scramjets, excess heat addition can result in a normal shock moving upstream from the combustor into the intake section (known as an inlet interaction) or eventually even moving out and standing in front of the intake. The additional loss of total pressure due to the normal shock can be quite high in such cases. These undesirable effects can be avoided altogether by choosing the second option for changing the slope of the Rayleigh line, namely, increasing the cross-sectional area.[2] For a given mass flow rate, increasing the cross-sectional area effectively increases q^*, since the heat release now occurs over a larger volume.

4.4 Calculation Procedure

The objective in any problem involving heat interaction is to calculate the final state, given an initial state and the amount of heat added/removed. Rather than solving the governing equations listed at the beginning of this chapter, it is easier to relate any state on the Rayleigh curve to the sonic state. Once this is done, the solution process becomes simple. We start with

$$\frac{dT_0}{T_0} = \frac{2(1 - M^2)}{1 + \gamma M^2} \frac{1}{1 + \frac{\gamma - 1}{2}M^2} \frac{dM}{M}$$

in Eq. 4.9. We can integrate this equation between any state and the sonic state to get

$$\frac{T_0}{T_0^*} = \frac{2(\gamma + 1)M^2 \left(1 + \frac{\gamma - 1}{2}M^2\right)}{(1 + \gamma M^2)^2} \tag{4.13}$$

T_0^* can be evaluated from this equation with the given M_1 and $T_{0,1}$. Since state point 2 lies on the same Rayleigh curve, T_0^* remains the same. From Eq. 4.2, we can get

---

[1] The flow at the exit of the duct is said to be under-expanded in this case. This is explained in detail in Chap. 6.

[2] The cross-sectional area of the entire combustor has to be increased, as, otherwise, the flow will not be one-dimensional. This type of variable area combustor would obviously introduce a lot of mechanical complexity in the combustor and hence is impractical. In actual supersonic combustors, the cross-sectional area increases along the length of the combustor. The increasing area accelerates the supersonic flow, as will be shown in Chap. 6, and counteracts the deceleration due to heat addition, thereby delaying thermal choking.

$$\frac{T_{0,2}}{T_0^*} = \frac{T_{0,1}}{T_0^*} + \frac{q}{C_p T_0^*}.$$

Since q is known, $T_{0,2}$ can be evaluated. With $T_{0,2}/T_0^*$ known, M_2 can be evaluated from Eq. 4.13. Similarly, by starting from Eq. 4.11, we can obtain a relationship for P_0/P_0^* in terms of M. From this, we can successively evaluate P_0^* and then $P_{0,2}$. With $T_{0,2}$, $P_{0,2}$, and M_2 known, all the other properties at state 2 can be evaluated.

In the actual calculations, tabular forms of relationships such as Eq. 4.13 are used as illustrated next. Values of P/P^*, T/T^*, ρ/ρ^*, P_0/P_0^* and T_0/T_0^* are listed in Table A.3 for values of M_1 ranging from 0.01 to 5.

Example 4.1 Air ($\gamma = 1.4$, molecular weight $= 28.8$ kg/kmol), enters a combustion chamber at 69 m/s, 300 K and 150 kPa, where 900 kJ/kg of heat is added. Determine (a) the mass flow rate per unit duct area, (b) exit properties, and (c) inlet Mach number if the heat added is 1825 kJ/kg.

Solution : Given $P_1 = 150$ kPa, $T_1 = 300$ K, $u_1 = 69$ m/s.

$$M_1 = u_1/\sqrt{\gamma R T_1} = 0.2$$

$$T_{0,1} = T_1 \left(1 + \frac{\gamma - 1}{2} M_1^2\right) = 302 \text{ K}$$

and

$$P_{0,1} = P_1 \left(\frac{T_{0,1}}{T_1}\right)^{\frac{\gamma}{\gamma - 1}} = 154 \text{ kPa}$$

(a) $\dot{m} = \rho_1 u_1 A = \frac{P_1}{R T_1} u_1 A = \mathbf{0.12 \times 10^2}$ **kg/s**

(b) From the table for heat addition, for $M_1 = 0.2$, we get

$$\frac{T_{0,1}}{T_0^*} \approx 0.1736 \text{ and } \frac{P_{0,1}}{P_0^*} \approx 1.235$$

Therefore, $T_0^* = 1740$ K and $P_0^* = 125$ kPa. It follows that

$$\mathbf{T_{0,2}} = T_{0,1} + \frac{q}{C_p} = \mathbf{1193 \text{ K}}$$

Since $T_{0,2}/T_0^* = 0.6857$, from the table for heat addition, we get $M_2 \approx 0.49$ and $P_{0,2}/P_0^* = 1.118$. Hence $\mathbf{P_{0,2} = 140}$ **kPa**,

$$\mathbf{T_2} = \frac{T_{0,2}}{\left(1 + \frac{\gamma - 1}{2} M_2^2\right)} = \mathbf{1138 \text{ K}}$$

and

$$P_2 = \frac{P_{0,2}}{\left(1 + \frac{\gamma - 1}{2} M_2^2\right)^{\frac{\gamma}{\gamma - 1}}} = 119 \, \text{kPa}$$

(c) For the given inlet conditions, $q^* = C_p(T_0^* - T_{0,1}) = 1453$ kJ/kg. Since the heat to be added is greater than 1453 kJ/kg, we must find an inlet state $(1')$ for which $q^* = 1825$ kJ/kg. Inlet stagnation conditions remain the same. Hence,

$$T_0^* = \frac{q^*}{C_p} + T_{0,1'} = 2108 \, \text{K}$$

Thus,

$$\frac{T_{0,1'}}{T_0^*} = 0.1432 \quad \Rightarrow M_{1'} = 0.18$$

Problems

4.1 Air enters a constant area combustion chamber at 100 m/s and 400 K. Determine the exit conditions if the heat added is (a) 1000 kJ/kg and (b) 2000 kJ/kg. Also determine the ratio of mass flow rates between the two cases. Assume that the inlet stagnation conditions remain the same for the two cases.
    [1281 K, 0.67; 1941 K, 1; 1.233]

4.2 Air enters a combustion chamber at 75 m/s, 150 kPa and 300 K. Heat addition in the combustor amounts to 900 kJ/kg. Compute (a) the mass flow rate, (b) the exit properties and (c) amount of heat to be added to cause the exit Mach number to be unity.
    [130.7 kg/s; 0.6, 107 kPa, 1123 K; 1173 kJ]

4.3 Determine the inlet static conditions and the mass flow rate if the heat addition in the above combustor were 1400 kJ/kg.
    [0.205, 151 kPa, 300.3 K, 124 kg/s]

4.4 Air enters the combustor of a scramjet engine at $M = 2.5$ and $T_0 = 1500$ K and $P_0 = 1$ MPa. Kerosene with calorific value of 45 MJ/kg is used as the fuel. Determine the fuel–air ratio (on a mass basis) that will result in the exit Mach number being equal to 1. Also determine the fuel-air ratio that will result in an "inlet interaction" *i.e.,* a normal shock stands just at the entrance of the combustor.
    [0.01387, 0.01416]

4.5 Air flows from a large reservoir where the pressure and temperature are 200 kPa, 300 K respectively, through a pipe of diameter 0.05 m and exhausts into the atmosphere at 100 kPa. Heat is added to the air in the pipe.

If the heat added is 450 kJ/kg, calculate (a) the inlet and exit Mach numbers, (b) exit static pressure, and (c) mass flow rate through the pipe.

If the exit static pressure has to be equal to the ambient pressure, calculate (a) the inlet and exit Mach numbers, (b) maximum amount of heat that can be added, and (c) mass flow rate through the pipe.

[a) 0.33, 1.0 b) 89.06 kPa c) 0.4898 kg/s]

[a) 0.66, 1 b) 42 kJ/kg c) 0.812 kg/s]

Chapter 5
Flow with Friction—Fanno Flow

In this chapter, we look at 1D adiabatic flow in a duct with friction at the walls of the duct. This type of flow occurs, for example, when gases are transported through pipes over long distances. It is also of practical importance when equipment handling gases are connected to high-pressure reservoirs, which may be located some distance away. Knowledge of this flow will allow us to determine the mass flow rate that can be handled, pressure drop, and so on.

In a real flow, friction at the wall arises due to the viscosity of the fluid and this appears in the form of a shear stress at the wall. So far in our discussion, we have assumed the fluid to be calorically perfect and inviscid as well. Thus, strictly speaking, viscous effects cannot be accounted for in this formulation. However, in reality, viscous effects are confined to very thin regions ("boundary layers") near the walls. Effects such as viscous dissipation are also usually negligible. Hence, we can still assume the fluid to be inviscid and take the friction force exerted by the wall as an externally imposed force. The origin of this force is of no significance to the analysis.

5.1 Governing Equations

The governing equations for this flow are Eqs. 2.8, 2.10, and 2.12,

$$\rho_1 u_1 = \rho_2 u_2 \tag{2.8}$$

$$h_1 + \frac{u_1^2}{2} = h_2 + \frac{u_2^2}{2} \tag{2.10}$$

© The Author(s) 2021
V. Babu, *Fundamentals of Gas Dynamics*,
https://doi.org/10.1007/978-3-030-60819-4_5

$$s_2 - s_1 = C_v \ln \frac{P_2}{P_1} + C_p \ln \frac{v_2}{v_1} \qquad (2.12)$$

The momentum equation, Eq. 2.9 has to be modified to take into account frictional force at the wall and so

$$P_1 + \rho_1 u_1^2 = P_2 + \rho_2 u_2^2 + \frac{\mathcal{P}}{A} \int_0^L \tau_w dx$$

where $\mathcal{P}$ is the wetted perimeter, L is the length of the duct, and τ_w is the wall shear stress. The Darcy friction factor f is related to the wall shear stress as $f/4 = \tau_w / \frac{1}{2}\rho u^2$. Upon using this relationship, we can write the above equation as

$$P_1 + \rho_1 u_1^2 = P_2 + \rho_2 u_2^2 + \frac{4}{D_h} \int_0^L \frac{1}{2}\rho u^2 \frac{f}{4} dx \qquad (5.1)$$

where $D_h = 4A/\mathcal{P}$ is the hydraulic diameter. The friction factor f can be calculated from Moody's chart[1] and is usually assumed to be constant along the duct (or pipe).

5.2   Illustration on T-s Diagram

We follow the same procedure as in the previous chapter and try to determine the locus of the allowed downstream states, starting from a given inlet state. From Eq. 4.3

$$\frac{d\rho}{\rho} = -\frac{du}{u} \qquad (4.2)$$

From the definition of stagnation temperature,

$$dT_0 = dT + \frac{1}{C_p} u du$$

Since the flow is adiabatic, there is no change in stagnation temperature and so $dT_0 = 0$ and this equation can be simplified to read

---

[1] Alternatively, the Colebrook formula can be used for determining the Darcy friction factor. Here

$$\frac{1}{f^{1/2}} = -2 \log \left(\frac{\epsilon/D_h}{3.7} + \frac{2.51}{Re\, f^{1/2}} \right)$$

where the Reynolds number is based on the hydraulic diameter and the mean velocity and ϵ is the roughness of the pipe surface.

$$\frac{dT}{T} = -(\gamma - 1)M^2 \frac{du}{u} \tag{5.2}$$

From the equation of state $P = \rho RT$, we get

$$dP = \rho R dT + RT d\rho$$

Substituting for dT and $d\rho$ from above and simplifying, we get

$$\frac{dP}{P} = -\left[1 + (\gamma - 1)M^2\right] \frac{du}{u} \tag{5.3}$$

Equation 2.6 can be written as

$$ds = C_v \frac{dP}{P} - C_p \frac{d\rho}{\rho}$$

If we substitute for dP and $d\rho$ from above, we get

$$ds = R(1 - M^2) \frac{du}{u} . \tag{5.4}$$

Finally, from the definition of Mach number, $M = u/\sqrt{\gamma RT}$, we can write

$$dM = M \frac{du}{u} - \frac{M}{2} \frac{dT}{T}$$

This can be simplified by using Eq. 5.2 to read

$$\frac{dM}{M} = \left(1 + \frac{\gamma - 1}{2} M^2\right) \frac{du}{u} \tag{5.5}$$

Since the flow is adiabatic and friction represents irreversibility, the entropy has to increase along the direction of flow. In other words, $ds > 0$ as we move from one state point to the next along the flow. Hence, it is more convenient to eliminate du/u using Eq. 5.4 in favor of ds in the above equations. This leads to

$$\frac{d\rho}{\rho} = -\frac{1}{R(1 - M^2)} ds$$

$$\frac{dT}{T} = -(\gamma - 1)M^2 \frac{1}{R(1 - M^2)} ds$$

$$\frac{dP}{P} = -\left[1 + (\gamma - 1)M^2\right] \frac{1}{R(1 - M^2)} ds \tag{5.6}$$

Table 5.1 Changes in properties along the flow direction

	$s \uparrow$
$M < 1$	$\rho \downarrow P \downarrow T \downarrow u \uparrow M \uparrow$
$M > 1$	$\rho \uparrow P \uparrow T \uparrow u \downarrow M \downarrow$

$$\frac{dM}{M} = \left(1 + \frac{\gamma - 1}{2}M^2\right) \frac{1}{R(1 - M^2)} \, ds$$

$$\frac{du}{u} = \frac{1}{R(1 - M^2)} \, ds$$

Let us now go ahead and summarize the changes in properties as we move from one state point to the next in the direction of flow.

The observations in Table 5.1 can be summarized conveniently as follows. The effect of friction on a subsonic flow is to increase the velocity, Mach number and decrease the static temperature and static pressure. Thus, the next state point lies to the right and below at a lower static pressure and temperature on the T-s diagram. On the other hand, the effect of friction on a supersonic flow, is to increase the static temperature and static pressure, while velocity and Mach number decrease. Thus, the next state point lies to the right and above at a higher static pressure and temperature on the T-s diagram.

These findings allow us to construct the locus of all the possible states (for the given inlet state or mass flow rate) and ultimately the state at the end of the duct, step by step. This curve is called the Fanno curve and is illustrated in Fig. 5.1. Furthermore, by combining Eqs. 5.2 and 5.4, we can get

$$\frac{dT}{ds} = -\frac{M^2}{1 - M^2} \frac{T}{C_v} \tag{5.7}$$

The following inferences may be drawn from Eq. 5.7:

- The slope of the subsonic portion of the Fanno curve is negative, while the slope of the supersonic portion is positive.
- A comparison of Eq. 5.7 with Eqs. 2.21 and 2.22 shows that the supersonic branch of the Fanno curve is steeper than the isochor and isobar.
- $dT/ds \to \mp\infty$ as $M \to 1$ and the sonic state occurs at the point of maximum entropy like the Rayleigh curve. However, unlike the Rayleigh curve, it is not possible to move through the sonic state on a Fanno curve.

Since friction renders the process irreversible, the stagnation pressure always decreases in a Fanno flow. This can be seen from Eq. 2.19, after noting that $T_{0,2} = T_{0,1}$ and $s_2 > s_1$. Alternatively, we can follow the steps used in the previous chapter and show that

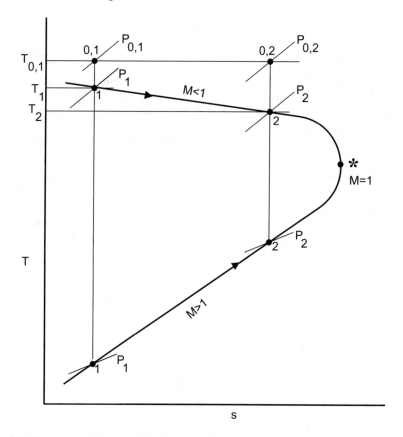

Fig. 5.1 Illustration of Flow with Friction on T-s diagram

$$\frac{dP_0}{P_0} = \frac{dP}{P} - \frac{\gamma}{\gamma - 1}\frac{dT}{T}$$

If we substitute from Eqs. 5.2 and 5.3, and then from Eq. 5.5, we get

$$\frac{dP_0}{P_0} = -\frac{ds}{R} = \frac{1 - M^2}{1 + \frac{\gamma - 1}{2}M^2}\frac{dM}{M} \qquad (5.8)$$

The first equality makes it clear that there is always a loss of stagnation pressure since $ds > 0$ regardless of the Mach number.

5.3   Friction Choking and Its Consequences

It is clear from Fig. 5.1 that, for a given initial state 1 at the entrance of the duct, there is a certain duct length L^* for which the exit state is the sonic state. For this duct length, the flow is choked at the exit. Since this choking is a consequence of friction, it is called *friction choking*. Similar to flow with heat addition, we wish to find out what would happen if the length of the duct were greater than L^*. Not surprisingly, the answer to this question is the same as that for flow with heat addition—in the case of subsonic flow, the inlet static conditions are changed so as to have a reduced mass flow rate (duct area being the same) and in the case of supersonic flow, a normal shock stands somewhere in the duct. The resulting Fanno process is shown in Figs. 5.2a, b on a T-s diagram. Although the exit state is shown to be the sonic state in this figure, the actual exit state will be such that the exit static pressure matches the ambient pressure in the region into which the pipe exhausts and the exit Mach number will be less than unity. However, if the value of the ambient pressure is too low, then, the exit Mach number will be equal to unity with the exit static pressure being more than the ambient value.[2]   Increasing the area of the duct effectively increases L^*. This makes sense since the pressure drop due to frictional effect decreases with increasing cross-sectional area or diameter (see Eq. 5.1) (Fig. 5.2).

Fig. 5.2   a Illustration of Fanno process with duct length $L > L^*$. Subsonic inlet

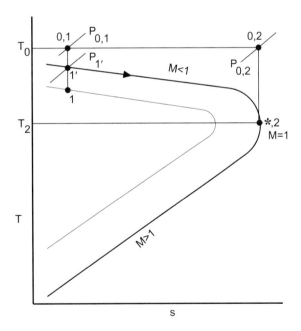

---

[2]The flow at the exit of the pipe is said to be under-expanded in this case. This is explained in detail in Chap. 6.

Fig. 5.2 b Illustration of
Fanno process with duct
length $L > L^*$. Supersonic
inlet

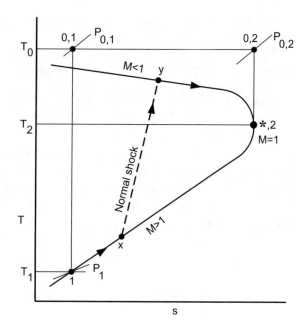

5.4 Calculation Procedure

For calculations involving Fanno flow, we will use the same procedure as what we
used for Rayleigh flow. That is, relate any state on the Fanno curve to the sonic state.
To do this, we start by writing Eq. 5.1 in differential form:

$$dP + \rho u\,du + \frac{4}{D_h}\frac{1}{2}\rho u^2 \frac{f}{4}\,dx = 0$$

where the continuity equation has been used to set $d(\rho u) = \rho u\,du$. By using the
equation of state and the definition of the Mach number, this can be further simplified
as

$$\frac{dP}{P} + \gamma M^2 \frac{du}{u} + \frac{4}{D_h}\frac{f}{4}\frac{\gamma M^2}{2}\,dx = 0$$

Substituting for dP/P from Eq. 5.3 in terms of du/u and then eliminating du/u
in favor of dM/M using Eq. 5.5 leads to

$$\frac{M^2 - 1}{\gamma M^2}\frac{1}{1 + \frac{\gamma - 1}{2}M^2}\frac{dM^2}{M^2} = -\frac{4}{D_h}\frac{f}{4}\,dx$$

This equation can be integrated between any state and the sonic state to obtain L^*. For a given inlet state 1, we can thus evaluate L_1^*. If the length of the duct is L, then since both states 1 and 2 (exit of the duct) lie on the same Fanno curve, $L_2^* = L_1^* - L$. With L_2^* known, M_2 can be evaluated from the above equation. Once M_2 is known, all the other properties at the exit can be calculated. As usual, instead of trying to work with complex closed form analytical expressions, we will use gas tables for the analysis. Table A.4 lists P/P^*, T/T^*, ρ/ρ^*, P_0/P_0^* and $f L^*/D_h$ for values of M_1 from 0.01 to 5.

Example 5.1 Air $(\gamma = 1.4$, molecular weight $= 28.8$ kg/kmol) enters a 3 cm diameter pipe with stagnation pressure and temperature of 100 kPa and 300 K and velocity of 100 m/s. Compute (a) the mass flow rate, (b) the maximum pipe length for this mass flow rate, and (c) mass flow rate for a pipe length of 14.5 m. Take $f = 0.02$.

Solution From the given data, we can get

$$T_1 = T_{0,1} - \frac{u_1^2}{2C_p} = 295\,\text{K}$$

$$M_1 = \frac{u_1}{\sqrt{\gamma R T_1}} = 0.29$$

and

$$P_1 = \frac{P_{0,1}}{\left(T_{0,1}/T_1\right)^{\gamma/\gamma-1}} = 94\,\text{kPa}$$

(a) Therefore

$$\dot{m} = \frac{P_1}{R T_1} A u_1 = \mathbf{0.078\,kg/s}$$

(b) From the gas tables, for $M_1 = 0.29$, we can get $f L^*/D = 5.79891$. Thus, $L^* = \mathbf{8.698\,m}$

Also, for this length $P_{0,1}/P_0^* = 2.1$, which represents an almost 52 percent loss of stagnation pressure at the pipe exit.

(c) Since the given length is greater than the L^* for this inlet condition, we have to determine the inlet Mach number for which L^* is the same as the given length. From the tables, for $f L/D = 9.6667$, this comes out to be $M_{1'} \approx 0.24$. Hence

$$T_{1'} = \frac{T_{0,1}}{\left(1 + \frac{\gamma - 1}{2} M_{1'}^2\right)} = 296.6\,\text{K}$$

$$P_{1'} = \frac{P_{0,1}}{\left(\frac{T_{0,1}}{T_{1'}}\right)^{\frac{\gamma}{\gamma - 1}}} = 96.1\,\text{kPa}$$

$$u_{1'} = M_{1'} * \sqrt{\gamma R T_{1'}} = 83.1\,\text{m/s}$$

and

$$\dot{m} = \frac{P_1}{R} T_1 \, A u_1 = 0.066 \, \text{kg/s}$$

Note the 15.4 percent reduction in the mass flow rate.

Example 5.2 Air ($\gamma = 1.4$, molecular weight $= 28.8$ kg/kmol) enters a 5 cm by 5 cm square duct at 300 K, 100 kPa and a velocity of 905 m/s. If the duct length is 2 m, find the flow properties at the exit. Take $f = 0.02$.

Solution From the give data, we have $M_1 = u_1/\sqrt{\gamma R T_1} = 2.6$. From the gas tables, for $M_1 = 2.6$, we can get $f L_1^*/D = 0.4526$. Since the cross section of the duct is noncircular, we use the hydraulic diameter, $D_h = 4A/\mathcal{P} = 5$ cm to get $L_1^* = 1.1315$ m. The given duct length is greater than this length and so there will be a normal shock standing somewhere in the duct.

Since no information is given about the exit static pressure, we will assume the exit state to be the sonic state. The location of the normal shock has to be determined iteratively. Let the condition immediately upstream of the shock be denoted with subscript x, and condition immediately downstream with subscript y, and let the location of the shock from the inlet be denoted by L_s. With an assumed value for L_s during each iteration, the calculations proceed along each row from left to right as shown in the table. After 3 iterations, we can say that the normal shock is located at $L_s \approx 0.25\,m$ from the inlet.

Iteration No	L_s (m)	L_x^* (m) $= L_1^* - L_s$	M_x (from Fanno table)	M_y (from Normal shock table)	L_y^* (m) (from Fanno table)	$L - L_s$ (m)
1	1	0.1315	≈ 1.26	0.8071	0.166	1
2	0.5	0.6315	≈ 1.84	0.6078	1.155	1.5
3	0.25	0.8815	2.1686	0.5537	1.77	1.75

Flow properties just before the shock (from Fanno table) are

$$T_x = \left(\frac{T_x}{T_1^*}\right) \left(\frac{T_1^*}{T_1}\right) T_1 = \frac{0.6235}{0.5102} \times 300 = 366.62 \text{ K}$$

$$P_x = \left(\frac{P_x}{P_1^*}\right) \left(\frac{P_1^*}{P_1}\right) P_1 = \frac{0.3673}{0.2747} \times 100 = 133.71 \text{ kPa}$$

and

$$P_{0,x} = \left(\frac{P_{0,x}}{P_{0,1}^*}\right) \left(\frac{P_{0,1}^*}{P_{0,1}}\right) \left(\frac{P_{0,1}}{P_1}\right) P_1$$

$$= \frac{1.919}{2.896} \times 19.95 * \times 100 = 1322 \text{ kPa}$$

Flow properties just after the shock (from normal shock table) are

$$T_y = \frac{T_y}{T_x} T_x = 1.813 \times 366.62 = 664.68 \text{ K}$$

$$P_y = \frac{P_y}{P_x} P_x = 5.226 \times 133.71 = 699 \text{ kPa}$$

$$P_{0,y} = \frac{P_{0,y}}{P_{0,x}} P_{0,x} = 0.6511 \times 1322 = 860.75 \text{ kPa}$$

Flow properties at the exit (from Fanno table) are

$$\mathbf{T_2} = T_2^* = \frac{T_2^*}{T_y} T_y = \frac{664.68}{1.1305} = \mathbf{588 \, K}$$

$$\mathbf{P_2} = P_2^* = \frac{P_2^*}{P_y} P_y = \frac{699}{1.9202} = \mathbf{364 \, kPa}$$

$$\mathbf{P_{0,2}} = P_{0,2}^* = \frac{P_{0,2}^*}{P_{0,y}} P_{0,y} = \frac{860.75}{1.249} = \mathbf{689 \, kPa}$$

$$\mathbf{T_{0,2}} = T_{0,1} = \frac{T_{0,1}}{T_1} T_1 = 2.352 \times 300 = \mathbf{705.6 \, K}$$

Note the 65 percent loss of stagnation pressure at the duct exit.

Problems

5.1 Redraw Fig. 5.1 in P-v coordinates.

5.2 Air enters a 5 cm × 5 cm smooth, insulated square duct with a velocity of 900 m/s, and a static temperature of 300 K. If the duct length is 2 m, determine the flow conditions at the exit. Use Colebrook's formula to calculate the friction factor.
[1.3, 528 K]

5.3 Air enters a smooth, insulated circular duct at $M = 3$. Determine the stagnation pressure loss in the duct for L/D = 20 and 40. Take $f = 0.02$.
[72.72%, 76.4%]

5.4 Air enters a smooth, insulated 3 cm diameter duct with stagnation pressure and temperature of 200 kPa, 500 K, and a velocity of 100 m/s. Compute (a) the maximum duct length for these conditions, (b) the mass flow rate for a duct length of 15 m and 30 m. Take $f = 0.02$.

[16.5 m; 0.097 kg/s, 0.076 kg/s]

5.5 Air enters a 3 m long pipe ($f = 0.02$) of diameter 0.025 m at a stagnation temperature of 300 K. If the static pressure of the air at the exit of the pipe is 100 kPa and the Mach number is 0.7, determine the stagnation pressure at entry and the mass flow rate through the pipe. [207.73 kPa, 0.14518 kg/s]

5.6 Air enters a pipe ($f = 0.02$) of diameter 0.05 m with stagnation pressure and temperature equal to 1 MPa and 300 K, respectively. The pipe exhausts into the ambient at 100 kPa. Determine the length of the pipe required to achieve a mass flow rate of 2 kg/s. Assume the flow to be subsonic at the entrance of the pipe.

[18.2 m]

Chapter 6
Quasi One Dimensional Flows

In the previous chapters, one-dimensional compressible flow solutions were presented, wherein the flow was either across a wave or through a constant area passage. A very important class of compressible flow is flow through a passage of finite but varying cross-sectional areas such as flow-through nozzles, diffusers, and blade passages in turbomachines. The main difficulty that arises in this case is that the flow is no longer strictly one-dimensional, since the variation in the cross-sectional area occurs in a direction normal to the main flow direction (see Fig. 6.1). This means that the velocity component in the normal direction is nonzero. However, it so happens that in most of the applications involving such flows, the magnitude of the normal component of velocity is *small* when compared with the axial component. Hence, as a first approximation, the former is usually neglected and only the axial component is considered. Thus, the flow is *approximately* one-dimensional, or, as it is usually called, quasi-one-dimensional.

Although it is possible to combine effects considered in previous chapters, such as friction and heat addition/removal with area variation, the resulting formulation is too complex to be considered in introductory texts such as the present one. The interested reader is referred to the books suggested at the end of the book for such an analysis. Here, we assume the flow through a varying area passage to be isentropic (except across normal shocks).

6.1 Governing Equations

The equations governing the flow are almost the same as those for one-dimensional flow and these are given below:

$$\rho_1 u_1 A_1 = \rho_2 u_2 A_2 \tag{6.1}$$

© The Author(s) 2021
V. Babu, *Fundamentals of Gas Dynamics*,
https://doi.org/10.1007/978-3-030-60819-4_6

Fig. 6.1 Flow through a
varying area passage

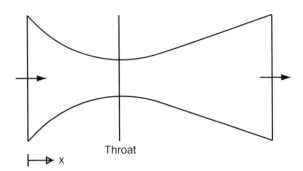

$$\left(P_1 + \rho_1 u_1^2\right) A_1 + \int_1^2 (P dA)_x = \left(P_2 + \rho_2 u_2^2\right) A_2 \tag{6.2}$$

$$h_1 + \frac{u_1^2}{2} = h_2 + \frac{u_2^2}{2} \tag{2.10}$$

Since the flow is isentropic, entropy remains the same, $s_2 = s_1$. Note that the cross-sectional area appears in the continuity and momentum equation.

6.1.1 Impulse Function and Thrust

The integral term in the momentum equation arises due to the pressure force on the wall. If the geometry under consideration is a nozzle such as in propulsion applications, this would be force exerted on the nozzle[1] and hence the airframe. To facilitate the evaluation of this force, we define a quantity called Impulse Function, I at any x-location as follows:

$$I = (P + \rho u^2)A \tag{6.3}$$

The net force exerted is the difference between the value of the impulse function at the exit and the inlet, viz.,

$$\mathcal{T} = I_2 - I_1$$

It is easy to see from this equation and Eq. 6.2 that $\mathcal{T} = \int_1^2 P dA$. Let us take the reference to propulsion application further and say that we are considering an aircraft

---

[1]It should be remembered that thrust force acts in the negative x-direction and drag force in the positive x-direction. Hence, a negative value for this integral would imply thrust and a positive value, drag.

flying at a speed of u_∞. From the above equation, the thrust produced by the engine (in a frame of reference where the aircraft is stationary and the flow approaches with a velocity of u_∞) is

$$\mathcal{T} = \left(PA + \rho u^2 A\right)_{exit} - \left(PA + \rho u^2 A\right)_{inlet}$$

If we use subscript ∞ to denote freestream conditions, and subscript e to denote exit conditions, then

$$\mathcal{T} = (PA + \dot{m}u)_e - (PA + \dot{m}u)_\infty$$

where we have used $\dot{m} = \rho u A$. If the static pressures in this equation are measured relative to the freestream pressure, then the thrust is given by

$$\mathcal{T} = \dot{m}\left(u_e - u_\infty\right) + (P_e - P_\infty)\, A_e \tag{6.4}$$

The first term in this equation is called *momentum thrust* and the second term is called *pressure thrust* and is nonzero when the pressure of the fluid at the exit is not equal to the ambient pressure. The negative term, $-\dot{m}u_\infty$, is called *intake momentum drag*. Note that in the case of a rocket engine, air is not taken in through any inlet and so this term will be absent.

6.2   Area Velocity Relation

The general objective in quasi-one-dimensional flows is to determine the Mach number at any axial location, given the inlet conditions and the area at that location. Once the Mach number is known, all the other properties at that location can be determined from the inlet properties using the fact that the flow is isentropic. Before we do this, let us first explore the nature of the flow in detail. The continuity equation for this flow can be written in differential form as

$$d(\rho u A) = 0 \tag{6.5}$$

which simply says that the mass flow rate at any section $\rho u A$ is a constant. Momentum and energy equation in differential form are the same as Eqs. 2.2[2] and 2.3, *viz.*,

$$dP + \rho u\, du = 0 \tag{2.2}$$

---

[2]For a differential fluid element of width dx, the forces on the fluid element in the x-direction are: PA (acting the positive x-direction), $(P + dP)(A + dA) = PA + A\,dP + P\,dA$ acting in the negative x-direction and $P\frac{dA}{\sin\theta}\sin\theta = P\,dA$ in the positive x-direction. Here θ is the inclination of the wall to the horizontal at the given axial location and the last term is the force exerted by the wall *on* the fluid element. Therefore, the net force on the fluid element in the x-direction is $PA - PA - A\,dP - P\,dA + P\,dA = -A\,dP$.

$$dh + d\left(\frac{u^2}{2}\right) = 0 \tag{2.3}$$

If we compute the derivative in Eq. 6.5 using product rule and then divide by $\rho u A$, we get

$$\frac{d\rho}{\rho} + \frac{du}{u} + \frac{dA}{A} = 0 \tag{6.6}$$

At any point in the flow field, $T v^{\gamma-1} = T_0 v_0^{\gamma-1}$. This can be rewritten as $T^{1/(\gamma-1)}/\rho = T_0^{1/(\gamma-1)}/\rho_0$, since $v = 1/\rho$. If we take the logarithm of this expression and then differentiate, we get

$$\frac{d\rho}{\rho} = \frac{1}{\gamma - 1}\frac{dT}{T}$$

Here, we have used the fact that the stagnation quantities remain constant as the flow is isentropic. Since $T_0 = T + u^2/(2C_p)$, we can write

$$dT = -\frac{\gamma - 1}{\gamma R} u \, du$$

Upon dividing both sides by T and rearranging, we get

$$\frac{1}{\gamma - 1}\frac{dT}{T} = -\frac{1}{\gamma R T}u^2\frac{du}{u}$$

which gives[3]

$$\frac{d\rho}{\rho} = -M^2\frac{du}{u} \tag{6.7}$$

If we substitute this into Eq. 6.6, we get

$$\frac{dA}{A} = (M^2 - 1)\frac{du}{u} \tag{6.8}$$

which is called the area–velocity relationship. The change in velocity for a given change in the area predicted by this equation for subsonic and supersonic flow is given in Table 6.1. It can be seen that a subsonic flow decelerates in a diverging passage

---

[3] Alternatively, we can write

$$\frac{d\rho}{\rho} = \frac{d\rho}{dP}\frac{dP}{\rho}$$

Since $dP/d\rho = a^2$, from Eq. 2.13 and $dP = -\rho u du$ from Eq. 2.2, we can write

$$\frac{d\rho}{\rho} = -M^2\frac{du}{u}$$

Table 6.1 Changes in velocity for a given change in area

	$A \uparrow$	$A \downarrow$
$M < 1$	$u \downarrow$	$u \uparrow$
$M > 1$	$u \uparrow$	$u \downarrow$

and accelerates in a converging passage. In contrast, a supersonic flow accelerates in a diverging passage and decelerates in a converging passage. This conclusion can be reached through a slightly different argument as follows.

We already know from Sect. 2.4 that changes in v for a given change in T are higher at lower values of temperature. This qualitative statement is made more precise in Eq. 6.7. Changes in density for a given change in velocity are higher when the flow is supersonic than when the flow is subsonic. Let us say that the velocity at a point in a subsonic flow increases by du. From Eq. 6.7, $d\rho$ is negative and since $M < 1$, $d\rho/\rho$ is less than du/u. It follows from Eq. 6.6 that dA has to be negative to make the left-hand side zero. Hence, if a subsonic flow accelerates, it can do so only in a converging passage. Similarly, let us say that the velocity at a point in a supersonic flow increases by du. From Eq. 6.7, $d\rho$ is negative and since $M > 1$, $d\rho/\rho$ is greater than du/u in magnitude. It follows from Eq. 6.6 that dA has to be positive to make the left-hand side zero, leading to the conclusion that a supersonic flow can accelerate only in a diverging passage. It is easy to demonstrate the remaining observations in Table 6.1 using similar arguments.

6.3 Geometric Choking

It is easy to see from Eq. 6.8 that, as $M \to 1$ on the right-hand side, then $dA \to 0$ on the left-hand side for the velocity to remain finite. In fact, $dA = 0$ where $M = 1$. Expressed another way, in an isentropic flow in a passage of varying cross sections, the sonic state can be attained only in a location where $dA = 0$. This location can be either a minimum (throat) or a maximum in the cross-sectional area as illustrated in Fig. 6.2. We have to determine whether it is possible to have $M = 1$ in both cases.

Let us consider the geometry shown on top in Fig. 6.2. If we assume the flow at the inlet to be subsonic, this will accelerate in the converging passage and can attain $M = 1$ at the throat. If, on the other hand, the flow is supersonic at the inlet, then this will decelerate in the converging passage and can possibly reach $M = 1$ at the throat. On the contrary, for the geometry shown in the bottom in Fig. 6.2, if we start with a subsonic flow at the inlet, it will decelerate in the diverging passage and so cannot attain $M = 1$ at the location where $dA = 0$. Similarly, if we start with a supersonic flow at the inlet, then since the supersonic flow accelerates in a diverging passage, it is not possible to reach $M = 1$ at the location where $dA = 0$. Thus, it is clear that an isentropic flow in a varying area passage can attain $M = 1$ only at a throat (minimum area of cross section).

Fig. 6.2 Illustration of
geometric choking

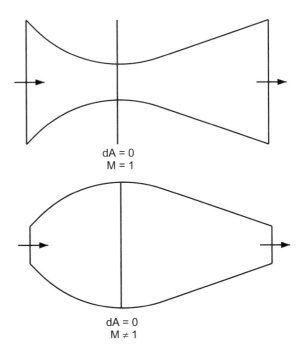

However, it is very important to realize that the converse need not be always true
i.e., the Mach number does not always have to be unity at a throat. This can be
established from Eq. 6.8 as follows. If we allow $dA \rightarrow 0$ on the left-hand side of
Eq. 6.8, then either $M \rightarrow 1$ or $du \rightarrow 0$ on the right-hand side, since the velocity
itself is finite. Of these two possibilities, the one realized in practice depends upon
the prevailing operating condition. Hence, for the geometry on the top in Fig. 6.2,
a flow, starting from a subsonic Mach number at the inlet can accelerate to a higher
Mach number (but less than one) at the throat and decelerate afterward. Similarly,
the flow can start from a supersonic Mach number, decelerate to a value above 1 at
the throat and accelerate again. The same argument is applicable to the geometry
in the bottom in Fig. 6.2. Two things should thus become clear from the arguments
given so far:

- Valid compressible flows are possible in both the geometries shown in Fig. 6.2. But,
  choking, if it occurs, can occur only in a geometric throat. This is a consequence
  of the fact that, in Eq. 6.8, when $M \rightarrow 1$, $dA \rightarrow 0$ for the velocity to be finite.
- Flow need not always choke at a geometric throat. This follows from the fact that,
  in Eq. 6.8, when $dA \rightarrow 0$ at the throat, $du \rightarrow 0$, without any restriction on the
  value of M.

When the Mach number does become equal to 1 at the throat, the flow is said to be choked and this choking is a consequence of the area variation and is called *geometric choking*.[4]

6.4 Area Mach Number Relation for Choked Flow

In a manner similar to the Rayleigh and Fanno flow problems, in the present case also, the state at any point can be conveniently related to the sonic state. We proceed to derive a relationship between the Mach number and cross-sectional area at any axial location to the area at the location where the sonic state occurs. We start by equating the mass flow rates at these two locations

$$\dot{m} = \rho u A = \rho^* u^* A^*$$

We can write

$$\frac{A}{A^*} = \frac{\rho^*}{\rho} \frac{u^*}{u} = \frac{\rho^*}{\rho_0} \frac{\rho_0}{\rho} \frac{u^*}{u}$$

where ρ_0 is the stagnation density. We know from Eq. 2.18 that

$$\frac{\rho_0}{\rho} = \left(1 + \frac{\gamma - 1}{2} M^2\right)^{\frac{1}{\gamma - 1}}$$

Setting $M = 1$ in this expression gives

$$\frac{\rho_0}{\rho^*} = \left(\frac{\gamma + 1}{2}\right)^{\frac{1}{\gamma - 1}}$$

Also, $u = Ma = M\sqrt{\gamma R T}$ and $u^* = a = \sqrt{\gamma R T^*}$. Substituting these expressions into the equation above for $\dot{m}$, we get

$$\frac{A}{A^*} = \left(1 + \frac{\gamma - 1}{2} M^2\right)^{\frac{1}{\gamma - 1}} \left(\frac{2}{\gamma + 1}\right)^{\frac{1}{\gamma - 1}} \frac{1}{M} \sqrt{\frac{T^*}{T}}$$

We can write T^*/T in terms of Mach number as follows.

---

[4]Mathematically, it is possible to have $M = 1$ at a location where $dA/dx = 0$ and $d^2A/dx^2 > 0$. These conditions are satisfied at locations where the area reaches a local minimum. Thus, we can have multiple locations where these conditions are satisfied as in Figs. 6.7 and 6.11. However, the Mach number will be equal to one at one or more of these locations depending upon the actual flow conditions.

$$\frac{T^*}{T} = \frac{T^*}{T_0}\frac{T_0}{T} = \frac{2}{\gamma + 1}\left(1 + \frac{\gamma - 1}{2}M^2\right)$$

If we substitute this into the equation above and simplify, we can finally write

$$\left(\frac{A}{A^*}\right)^2 = \frac{1}{M^2}\left[\frac{2}{\gamma + 1}\left(1 + \frac{\gamma - 1}{2}M^2\right)\right]^{\frac{\gamma + 1}{\gamma - 1}} \tag{6.9}$$

This relationship is called the Area Mach number relationship for an isentropic flow in a varying area passage. Given A, the area of cross section at a location and A^*, we can determine the Mach number at that location using this relation. Actually, this equation yields two solutions for a given A/A^*, one subsonic and the other supersonic, and the appropriate solution has to be chosen based on other details of the flow field. Also, it should be kept in mind that A^* is equal to the throat area, only when the flow is choked. Values of A/A^* for values of M ranging from 0.01 to 5 are listed in Table A.1.

6.5 Mass Flow Rate for Choked Flow

The mass flow rate at any section is given as

$$\dot{m} = \rho u A = \frac{\rho}{\rho_0}\rho_0 M\sqrt{\gamma R\frac{T}{T_0}T_0}\frac{A}{A^*}A^*$$

where we have used the same technique as in the previous section to rewrite the right-hand side. This can be rearranged to give

$$\dot{m} = \frac{P_0}{RT_0}\sqrt{\gamma RT_0}\,A_{\text{throat}}\frac{\rho}{\rho_0}M\sqrt{\frac{T}{T_0}}\frac{A}{A^*}$$

where we have used the equation of state $P_0 = \rho_0 RT_0$, and the fact that $A^* = A_{throat}$, since the flow is choked. Upon substituting for the last three terms in the right-hand side and after simplification, we are finally led to

$$\dot{m} = \frac{P_0 A_{\text{throat}}}{\sqrt{T_0}}\sqrt{\frac{\gamma}{R}\left(\frac{2}{\gamma + 1}\right)^{(\gamma + 1)/(\gamma - 1)}} \tag{6.10}$$

This equation is of tremendous importance in the design of intakes, nozzles, and wind tunnels. The most striking feature of this expression is that it does not involve any downstream property. The quantity under the big square root depends only upon the nature of the gas, such as whether it is monatomic or diatomic and the molecular weight. For a given working substance such as air, Eq. 6.10 shows that, once the flow is choked, the mass flow rate that can be realized through the passage is dependent only on the upstream stagnation pressure, temperature, and the throat area. This means that the mass flow rate cannot be controlled anymore from downstream, *i.e.,* by adjusting the exit conditions. In other words, this is the maximum mass flow rate that can be achieved by adjusting the backpressure. Also, any irreversibility upstream of the passage, which leads to a loss of stagnation pressure (such as normal shock or friction) reduces the mass flow rate for a given stagnation temperature and throat area. An increase of the upstream stagnation temperature (heat addition) also lowers the mass flow rate, but the reduction is more in this case since the stagnation pressure also decreases due to heat addition. It is clear from this expression that the mass flow rate can be changed at will, by an adjustment of the upstream stagnation conditions or the throat area. These are active control measures which can be utilized in practical devices when they operate under off-design conditions.

The quantity $P_0 A_{throat}/\dot{m}$ has units of velocity and is usually referred to as the characteristic velocity C^* in rocket propulsion. Thus,

$$C^* = \frac{P_0 A_{throat}}{\dot{m}} = \frac{1}{\Gamma} \sqrt{\frac{\mathcal{R} T_0}{\mathcal{M}}} \qquad (6.11)$$

where

$$\Gamma = \sqrt{\gamma \left(\frac{2}{\gamma + 1}\right)^{(\gamma + 1)/(\gamma - 1)}}$$

from Eq. 6.10, $\mathcal{R}$ is the Universal Gas Constant and $\mathcal{M}$ is the molecular weight of the gas.

6.6   Flow Through a Convergent Nozzle

Flow through a convergent nozzle can be established in one of the two ways:

- By *pulling* the flow—lowering the back pressure or the pressure of the ambient environment into which the nozzle exhausts, while maintaining the inlet stagnation conditions
- By *pushing* the flow—increasing the inlet stagnation pressure while maintaining the back pressure

The first scenario is illustrated using T-s coordinates in Fig. 6.3 and using P-v coordinates in Fig. 6.4. Initially, when the back pressure is less than P^* corresponding to the given inlet stagnation pressure P_0, the flow accelerates in the nozzle but the exit Mach number is less than one (Figs. 6.3a and 6.4a). The exit pressure of the fluid as it leaves the nozzle is the same as the ambient (back) pressure. It should be recalled that

Fig. 6.3 Illustration of flow through a convergent nozzle with inlet stagnation conditions fixed and varying back (ambient) pressure: T-s diagram

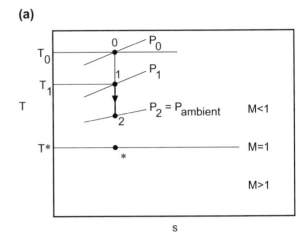

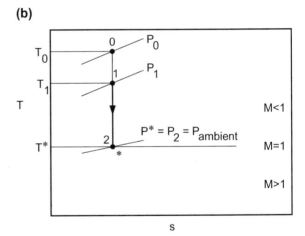

$$\frac{P_0}{P*} = \left(\frac{\gamma + 1}{2}\right)^{\frac{\gamma}{\gamma - 1}} = 1.8929 \, ,$$

and

$$\frac{T_0}{T*} = \frac{\gamma + 1}{2} = 1.2$$

where we have set $\gamma = 1.4$. When the back pressure is lowered to a value equal to $P*$, the flow accelerates and reaches a Mach number of one at the exit and the flow becomes choked. In this case also, the exit pressure of the fluid is the same as the ambient (back) pressure (Figs. 6.3b and 6.4b). Consequently, the diameter of the jet that issues out of the nozzle is exactly equal to the nozzle exit diameter. The mass

Fig. 6.4 Illustration of flow through a convergent nozzle with inlet stagnation conditions fixed and varying back (ambient) pressure: P-v diagram

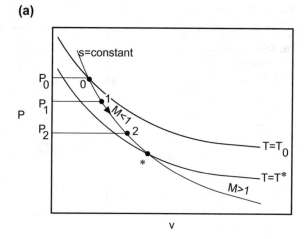

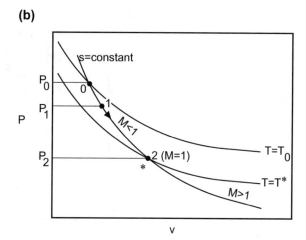

flow rate through the nozzle in this case is given by Eq. 6.10 and as discussed earlier, this is the maximum possible under these conditions.

If the back pressure is lowered further, the flow through the nozzle is unaltered. Physically, this is because the fluid is already traveling at the speed of sound at the exit and the changed back pressure condition is propagating *upstream* also at the speed of sound and so the flow becomes "aware" of the new back pressure value only after it reaches the exit. The static pressure of the fluid at the exit is still P^* but no longer equal to $P_{ambient}$. Since $P^* > P_{ambient}$ now, the fluid is "under-expanded" and it expands further outside the nozzle and equilibrates with the ambient conditions a few nozzle diameters downstream of the exit. The expansion is accomplished across an expansion fan (discussed in Chap. 8) centered at the nozzle lip. The jet swells initially as it comes out of the nozzle and expands, but shrinks afterward due to entrainment of the ambient air and equalization of static pressure.

Fig. 6.5 Illustration of flow
through a convergent nozzle
with varying inlet stagnation
conditions and fixed back
(ambient) pressure: T-s
diagram

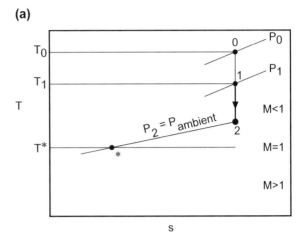

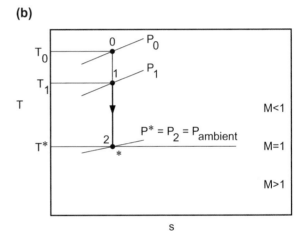

In propulsion applications, when the fluid expands outside the nozzle, the thrust
generated is less than optimum. Although it can be seen from Eq. 6.4 that the pressure thrust is nonzero in this case, this gain is more than offset by the reduction in
momentum thrust due to the reduced velocity of the fluid at the exit. If in a particular
propulsion nozzle this loss becomes too high, then the solution is to replace it with
a convergent divergent nozzle. This allows expansion of the fluid beyond the sonic
state inside the nozzle and hence recovers the lost thrust. It can be shown mathematically that the net thrust given by Eq. 6.4 is a maximum for a given $\dot{m}$, u_∞ and
P_∞, when the flow at the nozzle exit is correctly expanded. In order to show this, we
follow Zucrow & Hoffman and take the differential of Eq. 6.4 to get

$$dT = \dot{m}du_e + A_e dP_e + P_e dA_e - P_\infty dA_e$$

If we use the fact that $\dot{m} = \rho_e u_e A_e$ and rearrange the terms, we get

$$d\mathcal{T} = (\rho_e u_e du_e + dP_e) A_e + (P_e - P_\infty) dA_e$$

The term within the first bracket on the right-hand side can be seen to be zero from the differential form of the momentum equation, Eq. 2.2. Hence, it follows that $P_e = P_\infty$ for $d\mathcal{T} = 0$.

The second scenario, *i.e.*, pushing the flow is illustrated in Fig. 6.5 using T-s coordinates. When the stagnation pressure is not high enough, that is, $P_0/P_{\text{ambient}} < 1.8929$, then the flow is not choked at the exit. This is shown in Fig. 6.5a. As the stagnation pressure is increased (keeping the stagnation temperature constant), the exit Mach number and the mass flow rate both increase. The exit state point slides down along the $P = P_{\text{ambient}}$ isobar. When $P_0/P_{\text{ambient}} = 1.8929$, the flow becomes choked (Fig. 6.5b). Contrary to what happened in the previous scenario, if the stagnation pressure is increased further, then the mass flow rate also increases. However, the exit Mach number remains at 1. The exit static pressure is not equal to P_{ambient} any more but is equal to $P_0/1.8929$. Hence, the fluid is under-expanded and expands further outside the nozzle.

6.7 Flow Through a Convergent Divergent Nozzle

Convergent divergent nozzles are used in supersonic wind tunnels, turbomachinery and in propulsion applications such as aircraft engines and rockets. In propulsion applications, convergent nozzles can be used without severe penalty on the thrust upto $P_0/P_{\text{ambient}} < 3$. Beyond this value, convergent divergent nozzles have to be used to utilize the momentum thrust fully.

Flow through a convergent divergent nozzle also can be established in one of the two ways mentioned above. We look at the sequence of events during the start-up of a convergent-divergent nozzle with fixed inlet stagnation conditions and varying back pressure conditions, next. This sequence is illustrated in T-s coordinates using Fig. 6.6. The inlet and exit sections are denoted as before by 1 and 2. The corresponding variation of the static pressure along the length of the nozzle is shown in Fig. 6.7. Starting with Fig. 6.6a, we can see that, when the back pressure is high, the flow accelerates in the converging portion and decelerates in the diverging portion, but remains subsonic throughout (curve labeled (a) in Fig. 6.7). When the back pressure is reduced, the Mach number at the throat becomes 1 as shown in Fig. 6.6b and the flow becomes choked. The flow field from the inlet state to the throat as well as the mass flow rate through the nozzle does not change anymore (curve labeled (b) in Fig. 6.7).

When the back pressure is reduced some more, the flow accelerates beyond the throat and becomes supersonic (Fig. 6.6c). However, the back pressure is too high, and this triggers a normal shock in the divergent portion of the nozzle. The state points just before and after the shock are denoted by x and y, respectively, in Fig. 6.6.

Fig. 6.6 Illustration of flow
through a convergent
divergent nozzle with inlet
stagnation conditions fixed
and varying back (ambient)
pressure: T-s diagram

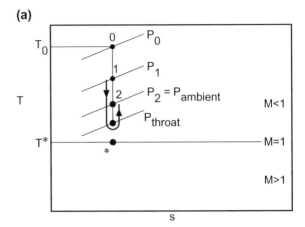

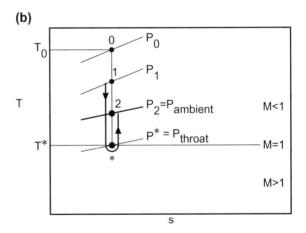

The flow becomes subsonic after the normal shock and it decelerates in the rest of the
divergent portion with the attendant increase in static pressure to the specified back
pressure (curve labeled (c) in Fig. 6.7). The location of the normal shock is dictated
by the exit area, throat area, and the back pressure. As the back pressure is lowered
further, the normal shock moves further downstream (Figs. 6.6d, e and curves labeled
(d) and (e) in Fig. 6.7). The situation shown in Fig. 6.6(e) where the normal shock
stands just at the exit represents a threshold situation. If the back pressure were to be
lowered further, then the normal shock moves out of the nozzle[5] and the flow inside
the nozzle becomes shock free as shown in Fig. 6.6f and the curve labeled (f) in
Fig. 6.7.

Two things should be noted in Fig. 6.6c–e. Firstly, the Mach number before the
shock keeps increasing as the shock moves downstream. Consequently, the loss of

---

[5]The normal shock actually becomes an oblique shock that is anchored to the nozzle lip.

Fig. 6.6 (continued)

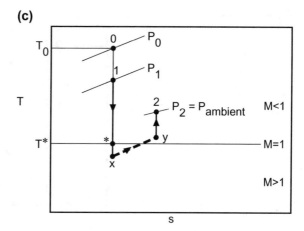

(c)

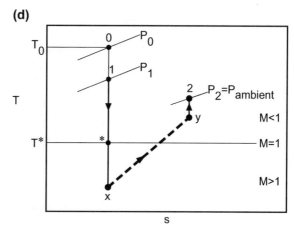

(d)

stagnation pressure across the shock wave also keeps increasing. Secondly, the flow field upstream of the shock wave does not change as the back pressure is lowered, since the flow is supersonic ahead of the shock wave. Finally, when the back pressure is decreased to the design value, the flow through the nozzle becomes shock free (isentropic) and the flow is supersonic throughout the divergent portion (Fig. 6.6f and curve labeled (f) in Fig. 6.7). Since the exit area and the throat area are known, and $M = 1$ at the throat, the exit Mach number can be calculated from Eq. 6.9.

If the back pressure is lowered below the design value, the nozzle exit pressure does not change. The jet is now said to be "under-expanded" (as in the case of the convergent nozzle) and further expansion takes place outside the nozzle.

In contrast to a convergent nozzle, it is possible to operate a convergent divergent nozzle in an "over-expanded" mode. This happens when the back pressure is higher than the design value, but lower than the value for which a normal shock would stand just at the exit (Fig. 6.7). Since the static pressure of the jet as it comes out of

Fig. 6.6 (continued)

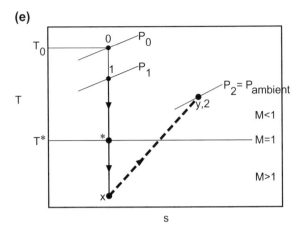

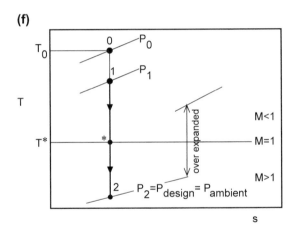

the nozzle is less than the ambient value, it undergoes compression through oblique shocks outside the nozzle.

The variation of thrust due to under-expanded and over-expanded mode of operation is a difficult issue with rocket engines, designed for a particular operational altitude. The jet is over-expanded at altitudes below the design altitude and under-expanded at higher altitudes. Altitude compensating nozzles are a better alternative under such circumstances.

The start-up sequence described above remains the same when the back pressure is fixed and the inlet stagnation pressure is varied. This means that the normal shock that occurs in the divergent portion is inevitable and cannot be avoided. This is undesirable since the loss of stagnation pressure across the normal shock can be quite high. It is for these reasons, that convergent divergent nozzles are not used in propulsion applications unless the pressure ratio $P_0/P_{ambient}$ is high enough.

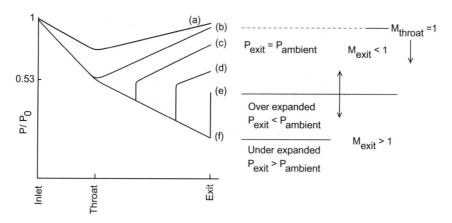

Fig. 6.7 Variation of static pressure in a convergent divergent nozzle with inlet stagnation conditions fixed and varying back (ambient) pressure. For condition (f), $P_{exit} = P_{design}$ and $M_{exit} = M_{design}$

Example 6.1 A converging diverging nozzle with an exit to throat ratio of 3.5, operates with inlet stagnation conditions 1 MPa and 500 K. Determine the exit conditions when the back pressure is (a) 20 kPa (b) 500 kPa. Assume air to be the working fluid ($\gamma = 1.4$, Mol. wt = 28.8 kg/kmol).

Solution Given $P_{0,1} = 1$ MPa, $T_0 = 500$ K and $A_{exit}/A_{throat} = 3.5$. For a correctly expanded flow, the exit conditions for this area ratio are (from the isentropic table)

$$\frac{P_{0,1}}{P_{exit}} = 27.14, \quad \frac{T_0}{T_{exit}} = 2.568 \quad \text{and } M_{exit} = 2.8$$

Therefore, $P_{exit} = 36.85$ kPa and $T_{exit} = 194.7$ K.

(a) Since the given exit static pressure of 20 kPa is less than the design value of 36.85 kPa, the flow is under-expanded. The values for the exit properties are the same as the design values.
(b) For a back pressure of 500 kPa, we have to see whether the flow is completely subsonic in the divergent portion or there is a normal shock. If we assume that the nozzle is still choked, then, for fully subsonic flow in the divergent portion, from the isentropic table, for an area ratio of 3.5,

$$1.01803 < \frac{P_{0,1}}{P_{exit}} < 1.02038 \Rightarrow 980\,kPa < P_{exit} < 982.29\,kPa$$

Since the given P_{exit} is less than this value, there is a normal shock in the divergent portion. The mass flow rate through the nozzle is the same at the throat and the exit sections. Thus,

$$\dot{m} = \rho^* u^* A_{throat} = \rho_{exit} u_{exit} A_{exit}$$

$$\Rightarrow \quad \frac{P^*}{RT^*} \sqrt{\gamma RT^*} \, A_{\text{throat}} = \frac{P_{\text{exit}}}{RT_{\text{exit}}} M_{\text{exit}} \sqrt{\gamma RT_{\text{exit}}} \, A_{\text{exit}}$$

$$\frac{P^*}{P_{0,1}} \frac{P_{0,1}}{P_{\text{exit}}} \sqrt{\frac{T_0}{T^*}} \frac{A_{\text{throat}}}{A_{\text{exit}}} = M_{\text{exit}} \sqrt{\frac{T_0}{T_{\text{exit}}}}$$

$$\Rightarrow \quad \frac{P^*}{P_{0,1}} \frac{P_{0,1}}{P_{\text{exit}}} \sqrt{\frac{\gamma+1}{2}} \frac{A_{\text{throat}}}{A_{\text{exit}}} = M_{\text{exit}} \sqrt{1 + \frac{\gamma-1}{2} M_{\text{exit}}^2}$$

$$\Rightarrow \quad \frac{1}{1.893} \frac{1000}{500} \sqrt{1.2} \frac{1}{3.5} = M_{\text{exit}} \sqrt{1 + 0.2 \, M_{\text{exit}}^2}$$

$$\Rightarrow \quad 0.2756 = M_{\text{exit}} \sqrt{1 + 0.2 \, M_{\text{exit}}^2}$$

Therefore, $\mathbf{M_{exit}} = \mathbf{0.2735}$,

$$\mathbf{T_{exit}} = \frac{T_{\text{exit}}}{T_0} T_0 = \frac{1}{1.015} 500 = \mathbf{493\,K}$$

and

$$\mathbf{P_{0,exit}} = \frac{P_0}{P_{exit}} P_{\text{exit}} = 1.056 \times 500 = \mathbf{528\,kPa}$$

The loss of stagnation pressure due to the normal shock is almost 50%.

Let x, y denote the states just ahead of and behind the normal shock. For $P_{0,y}/P_{0,x} = 528/1000 = 0.528$, from the normal shock table we get $M_x = 2.43$. From the isentropic table, A/A^* corresponding to this value of Mach number is 2.4714. Thus, the normal shock stands at a location in the divergent portion where A/A_{throat} is 2.4714.

Example 6.2 Air at 1100 kPa in a reservoir expands through a convergent divergent nozzle into the ambient. The exit to throat area ratio is 2. The ambient pressure is 120 kPa initially and is then gradually increased. Determine the Mach number and static and stagnation pressure at the exit as well as the ambient pressure at each of the following instants: (i) initial (ii) when a normal shock first appears (iii) when a normal shock occurs at a location where $A/A_{\text{throat}} = 1.55526$ and (iv) when the normal shock disappears.

Solution For shock-free flow with $M = 1$ at the throat (and so $A/A^* = 2$) and $M_e > 1$ (where the subscript e denotes the exit), we can get from Table A.1, $P_{0,e}/P_e = 10.6927$ and $M_e = 2.2$. Hence, $P_e = 1100 / 10.6927 = 103$ kPa. This corresponds to condition (f) in Fig. 6.7.

For shock-free flow with $M = 1$ at the throat and $M_e < 1$, we can get from Table A.1, $P_{0,e}/P_e = 1.06443$ and $M_e = 0.3$. Hence, $P_e = 1100 / 1.06443 = 1033.4$ kPa. This corresponds to condition (b) in Fig. 6.7.

For a normal shock at the exit, with x and y denoting the states just before and after the shock, it is easy to see that $M_x = 2.2$. From the normal shock table, for this value of M_x, we can retrieve, $M_y = 0.547$, $P_y/P_x = 5.48$, $P_{0,y}/P_{0,x} = 0.628$. This corresponds to condition (e) in Fig. 6.7. Note that state y corresponds to the exit state for this condition. Hence $M_e = 0.547$, $P_e = 5.48 \times 103 = 564.44$ kPa, $P_{0,e} = 0.628 \times 1100 = 690.8$ kPa.

(i) For this case, since $P_e < P_{ambient}$, the flow is over-expanded. Hence, $M_e = 2.2$, $P_e = 103$ kPa and $P_{0,e} = 1100$ kPa and $P_{ambient} = 120$ kPa.

(ii) A normal shock first appears when $P_{ambient} = 564.44$ kPa. Accordingly, $M_e = 0.547$, $P_{0,e} = 690.8$ kPa and $P_e = 564.44$ kPa.

(iii) As the ambient pressure is increased, the normal shock keeps moving upstream and ultimately disappears when it reaches the throat. Hence, $P_{ambient} = 1033.4$ kPa, $M_e = 0.3$, $P_{0,e} = 1100$ kPa and $P_e = 1033.4$ kPa.

(iv) It is given that $A_x/A_{throat} = 1.55526$. From the isentropic table, we can get $M_x = 1.9$. Next, from the normal shock table, for this value of M_x, we can get $M_y = 0.5956$ and $P_{0,y}/P_{0,x} = 0.767357$.

From the isentropic table, for $M = M_y = 0.5956$, we can get $A_y/A_y^* = 1.1935$. We can write

$$\frac{A_e}{A_y^*} = \frac{A_e}{A_{throat}} \frac{A_{throat}}{A_x} \frac{A_x}{A_y^*} = \frac{A_e}{A_{throat}} \frac{A_{throat}}{A_x} \frac{A_y}{A_y^*}$$

where we have used the fact that $A_y = A_x$, since the thickness of the normal shock is negligibly small. Hence, $A_e/A_y^* = 1.5348$. From the isentropic table, for this value of A/A^*, we can get $M_e \approx 0.418$ and $P_{0,e}/P_e = 1.127756$. Hence

$$P_e = \frac{P_e}{P_{0,e}} \frac{P_{0,y}}{P_{0,x}} P_{0,x} = \frac{0.767357 \times 1100}{1.127756} = 748.5 \text{ kPa}$$

where we have used the fact that $P_{0,e} = P_{0,y}$.

The occurrence of the normal shock during start-up of a convergent divergent nozzle leads to difficulties in non-propulsion applications also. Consider the supersonic wind tunnel shown in Fig. 6.8, where a convergent divergent nozzle is used to generate supersonic flow in the test section. In a continuously operating wind tunnel, the supersonic flow, after going through the test section, is usually diffused in a diffuser instead of being exhausted into the atmosphere. The supersonic diffuser is a mirror image of the supersonic nozzle. The diffuser allows the static pressure to be recovered and the flow can then be fed back to the nozzle. This results in enormous savings in energy required to run the tunnel. However, getting the tunnel "started" is difficult, due to the presence of the normal shock in the nozzle side combined with the second throat in the diffuser side. Here, starting means the establishment of shock-free operation with supersonic flow in the test section.

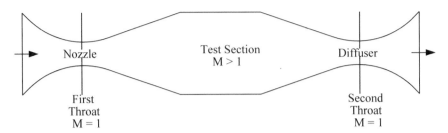

Fig. 6.8 Schematic illustration of a supersonic wind tunnel during steady state operation

During initial start-up, a normal shock stands in the divergent portion of the nozzle. At this point, the flow through the nozzle is choked and the mass flow rate through the nozzle, from Eq. 6.10, can be written as

$$\dot{m}_{\text{nozzle}} = \frac{P_{0,\text{nozzle}} A_{\text{throat,nozzle}}}{\sqrt{T_{0,\text{nozzle}}}} \sqrt{\frac{\gamma}{R} \left(\frac{2}{\gamma + 1}\right)^{(\gamma + 1)/(\gamma - 1)}}$$

If the diffuser also runs choked, then the mass flow rate through the diffuser can be written similarly as

$$\dot{m}_{\text{diffuser}} = \frac{P_{0,\text{diffuser}} A_{\text{throat,diffuser}}}{\sqrt{T_{0,\text{diffuser}}}} \sqrt{\frac{\gamma}{R} \left(\frac{2}{\gamma + 1}\right)^{(\gamma + 1)/(\gamma - 1)}}$$

There is no change in stagnation temperature between the nozzle and the diffuser section. However, because of the normal shock, there is a loss of stagnation pressure and so $P_{0,\text{diffuser}} < P_{0,\text{nozzle}}$. If the throat areas of the nozzle and the diffuser are the same, then the maximum mass flow rate that the diffuser can handle is less than the mass flow rate coming from the nozzle. The only way to accommodate the higher mass flow rate is to increase the diffuser throat area. If we equate the mass flow rates through the nozzle and diffuser, we get

$$\frac{A_{\text{throat,diffuser}}}{A_{\text{throat,nozzle}}} = \frac{P_{0,\text{nozzle}}}{P_{0,\text{diffuser}}}$$

The loss of stagnation pressure is the highest when the normal shock stands in the test section as the Mach number is the highest there. The diffuser throat should be sized for this condition. This allows the normal shock to be "swallowed" by the diffuser throat and the shock then moves into the divergent portion of the diffuser. It will eventually sit in the diffuser throat. At this point, the tunnel is shock free with supersonic flow in the test section and the tunnel is said to be "started". The diffuser throat area now has to be reduced. During steady-state operation, the diffuser throat area is equal to the nozzle throat area, although in reality, it will be larger.

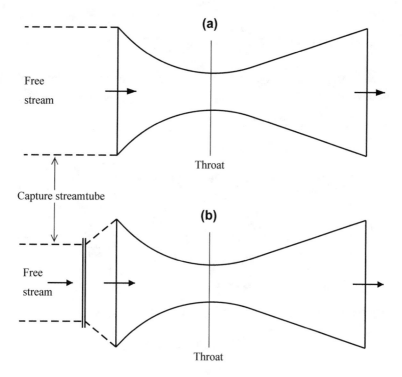

Fig. 6.9 Illustration of a supersonic diffuser at **a** design and **b** off-design operating condition

Example 6.3 Diffusers are commonly used in the intakes of supersonic vehicle to decelerate the incoming supersonic freestream to the appropriate Mach number before entry into the combustor. Consider a converging diverging diffuser in Fig. 6.9 (usually called an internal compression intake) that is designed for shock-free operation at a freestream Mach number of 2. Determine the ratio of the mass flow rate through the diffuser when the freestream Mach number is 1.5, to that at the design condition. Also determine the stagnation pressure recovery. Assume the intake geometry to be fixed and the throat Mach number to be 1 in all the cases. Freestream static conditions also remain the same.

Solution From the isentropic table, for $M_{inlet} = 2$, we get $A_{\text{inlet}}/A_{\text{throat}} = 1.6875$. For this value of $A_{\text{inlet}}/A_{\text{throat}}$, M_{inlet} can also be ≈ 0.37. During off-design operation with the throat Mach number at 1, and with the geometry fixed, the inlet Mach number *has to be* equal to this subsonic value. Since the Mach number at the inlet is subsonic, there has to be a normal shock (in reality, this will be a curved shock, but with a vertical portion which is a normal shock) standing in front of the diffuser. The shock stand-off distance will be different for different freestream Mach numbers, in such a way as to achieve this value of Mach number at the inlet. Thus, static conditions at the inlet will be different for different freestream Mach numbers. For operation at design Mach number,

$$\dot{m}_{\text{design}} = \frac{P_{0,\infty} A_{\text{throat}}}{\sqrt{T_{0,\infty}}} \sqrt{\frac{\gamma}{R} \left(\frac{2}{\gamma+1}\right)^{(\gamma+1)/(\gamma-1)}}$$

where the subscript ∞ denotes freestream condition. From isentropic table, for $M_\infty = 2$, we get

$$\frac{P_{0,\infty}}{P_\infty} = 7.82445 \text{ and } \frac{T_{0,\infty}}{T_\infty} = 1.8$$

$$\Rightarrow \dot{m}_{\text{design}} = 5.832 \frac{P_\infty A_{\text{throat}}}{\sqrt{T_\infty}} \sqrt{\frac{\gamma}{R} \left(\frac{2}{\gamma+1}\right)^{(\gamma+1)/(\gamma-1)}}$$

For operation with $M_\infty = 1.5$,

$$\dot{m}(M_\infty = 1.5) = \frac{P_{0,\text{inlet}} A_{\text{throat}}}{\sqrt{T_{0,\text{inlet}}}} \sqrt{\frac{\gamma}{R} \left(\frac{2}{\gamma+1}\right)^{(\gamma+1)/(\gamma-1)}}$$

From the normal shock table, for $M = 1.5$, we get

$$\frac{P_{0,\text{inlet}}}{P_{0,\infty}} = 0.929787$$

From the isentropic table, for $M_\infty = 1.5$, we get

$$\frac{P_{0,\infty}}{P_\infty} = 3.67103 \text{ and } \frac{T_{0,\infty}}{T_\infty} = 1.45$$

Therefore,

$$P_{0,\text{inlet}} = \frac{P_{0,\text{inlet}}}{P_{0,\infty}} \frac{P_{0,\infty}}{P_\infty} P_\infty = 3.4133 P_\infty$$

$$T_{0,\text{inlet}} = \frac{T_{0,\text{inlet}}}{T_{0,\infty}} \frac{T_{0,\infty}}{T_\infty} T_\infty = 1.45 T_\infty$$

and

$$\dot{m}(M_\infty = 1.5) = 2.8346 \frac{P_\infty A_{throat}}{\sqrt{T_\infty}} \sqrt{\frac{\gamma}{R} \left(\frac{2}{\gamma+1}\right)^{(\gamma+1)/(\gamma-1)}}$$

Therefore,

$$\frac{\dot{m}(M_\infty = 1.5)}{\dot{m}_{\text{design}}} = \mathbf{0.486}$$

and the total pressure recovery is 93%.

Shock-free operation even under off-design operating conditions can be achieved by having a variable area throat. The intake can be "started" by increasing the throat area such that

$$\frac{A_{\text{throat}}(M_\infty = 1.5)}{A_{\text{throat}}(\text{design})} = \frac{1}{0.486} = 2.058$$

During shock-free operation, the throat area should be reduced such that

$$\frac{A_{\text{throat}}(M_\infty = 1.5)}{A_{\text{throat}}(\text{design})} = \frac{1.6875}{1.17617} = 1.435$$

Thus, by adjusting the throat area to match the mass flow rate that the intake can "swallow", the normal shock in front of the intake under off-design operating conditions can be avoided. However, the required area variation is too large for such internal compression intakes to be of practical use.

Capture area of an intake is an important performance metric of supersonic intakes. The capture area is the freestream cross-sectional area of the streamtube that enters the intake. From Fig. 6.9, it is clear that the capture area for design operating condition is equal to the inlet area. For off-design operating condition, since the mass flow rate is less, the capture area also decreases. For the above example,

For operation with $M_\infty = 1.5$,

$$\frac{A_\infty}{A_{\text{throat}}} = \frac{2.8346}{M_\infty} \sqrt{\left(\frac{2}{\gamma + 1}\right)^{(\gamma+1)/(\gamma-1)}} = 1$$

For comparison, at design operating condition, $A_\infty / A_{\text{throat}} = 1.6875$

6.8 Interaction Between Nozzle Flow and Fanno, Rayleigh Flows

So far, we have looked at flow through a constant area passage with friction, with heat addition and isentropic flow through nozzles in isolation. In real-life applications, there will always be an interaction, since a nozzle has to be connected to a high-pressure reservoir located upstream through pipes, or heat addition in a combustion chamber is followed by expansion in a nozzle and so on. From Eq. 6.10, we know that the mass flow rate through a nozzle is affected by any upstream changes in stagnation pressure and temperature. Since Fanno flow and Rayleigh flow both result in changes in stagnation pressure (and stagnation temperature in the latter case), it is important to study the interaction between these flows.

Figure 6.10a shows a constant area passage followed by a convergent nozzle. In keeping with what we have discussed so far, we will assume frictional flow in the constant area passage and isentropic flow in the nozzle. The process is illustrated

Fig. 6.10 a Schematic illustration of a constant area passage followed by a convergent nozzle **b** Illustration of the flow on a T-s diagram

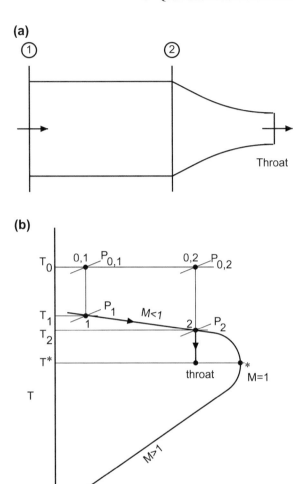

using T-s coordinates in Fig. 6.10b. State points 1 and 2 lie on the Fanno curve. The isentropic process in the nozzle is indicated by the vertical line going down to the sonic isotherm. Here, without any loss of generality, we have assumed the exit state to be sonic, but it need not be. As long as the exit Mach number is subsonic or just becoming sonic, the exit static pressure will be equal to the ambient pressure. Thus, it is possible to have geometric choking in this case, but not friction choking. Owing to the loss of stagnation pressure due to the effect of friction ahead of the nozzle, the choked mass flow rate through the nozzle is less now by a factor of $1 - P_{0,2}/P_{0,1}$. The longer the distance between the nozzle and the reservoir *i.e.,* longer the pipe connecting the two, the more the loss of stagnation pressure and lesser the mass flow rate. In fact, if the length of the pipe is greater than L^*, then, as we discussed in

Fig. 6.11 a Schematic illustration of a convergent nozzle followed by a constant area passage **b** Illustration of the flow on a T-s diagram

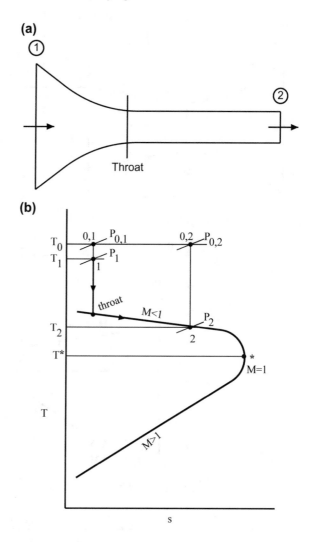

Sect. 5.3, the mass flow rate through the pipe will be reduced further. These factors must be borne in mind when designing equipment for handling compressible flow.

The situation when a convergent nozzle is located upstream of a constant area passage is illustrated in Fig. 6.11. Contrary to the previous case, now, the Mach number at the nozzle throat cannot reach 1, since the Mach number has to increase further in the constant area section due to friction. Hence, it is possible to have friction choking but not geometric choking in this case. Since the nozzle cannot choke, the mass flow rate is less than the maximum value possible for the given stagnation conditions at Sect. 1 and throat area. Thus, in this case also, the mass flow rate is reduced due to the effect of friction.

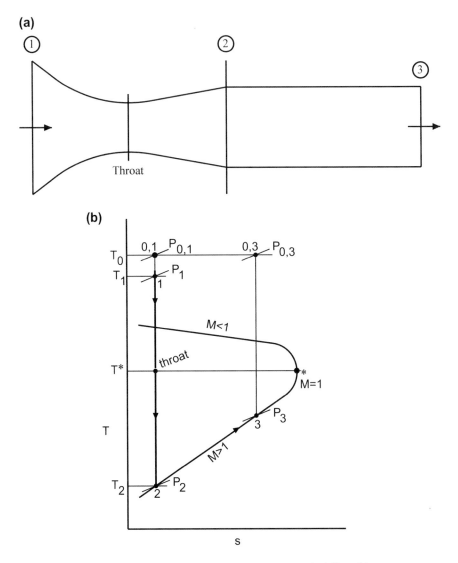

Fig. 6.12 a Schematic illustration of a convergent divergent nozzle followed by a constant area passage **b** Illustration of the flow on a T-s diagram $(L < L^*)$

Figure 6.12a shows a converging diverging nozzle feeding a constant area passage. The process is shown on a T-s diagram in Fig. 6.12b, when the length of the passage is less than L^* corresponding to the Mach number at the inlet of the passage. The isentropic expansion in the nozzle is denoted by the vertical line 1–2 in this figure, and the Fanno process in the constant area passage by 2–3 in the supersonic branch of the Fanno curve. If the length of the constant area passage is greater than the L^*

Fig. 6.12 (continued)

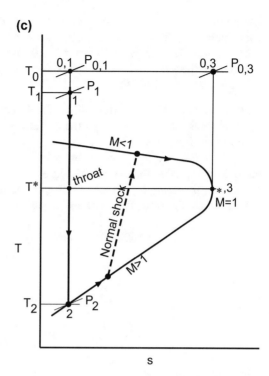

corresponding to M_2, then a normal shock occurs somewhere in the passage and the flow becomes subsonic. This is illustrated in Fig. 6.12c. There is no change in the mass flow rate, however. As the passage length is increased, this shock moves upstream and can even go into the divergent portion of the nozzle.

The interaction between flow with heat addition and isentropic flow in a nozzle can be developed along the same lines as discussed above. If the heat addition takes place before the nozzle, then the reduction in the mass flow rate is even higher, since heat addition not only increases the stagnation temperature but decreases the stagnation pressure as well. It is left as an exercise to the reader to illustrate the interaction on T-s diagrams similar to Figs. 6.10, 6.11, and 6.12. These considerations are important in practical propulsion applications such as combustors and afterburners. Most of the aircraft engines used in military applications have an afterburner for short duration thrust augmentation. The basic principle in an afterburner is to inject and burn fuel in the tailpipe portion located just before the nozzle. This increases the velocity of the fluid as it comes out of the nozzle, thereby increasing the thrust. Usually, a convergent nozzle is used with or without the afterburner in operation. However, the throat area of the nozzle has to be increased when the afterburner is in operation, if the mass flow rate is to be maintained, due to the above-mentioned factors. Consequently, such nozzles are made with interlocking flaps, which can be moved axially to control the exit (throat) area.

Example 6.4 Air ($\gamma = 1.4$, Mol. wt = 28.8 kg/kmol) flows through the nozzle-pipe combination shown in Fig. 6.12a. The stagnation conditions at the nozzle inlet are 1 MPa and 500 K. The pipe diameter is 0.05 m and it is 5 m long. Determine the reduction in mass flow rate due to the presence of the pipe. Take $f = 0.024$ and the back pressure to be 100 kPa.

Solution Given $P_{0,1} = 1$ MPa, $T_0 = 500$ K and $A_{throat} = \pi D^2 / 4 = 1.9625 \times 10^{-3}\, m^2$. Mass flow rate through the nozzle in the absence of the pipe is given by Eq. 6.10. Thus $\dot{m}(L = 0) = 3.54$ kg/s.

(a) For the nozzle-pipe combination, since the back pressure is given, we need to check to see whether the exit pressure is equal to or greater than the back pressure. We do this by assuming the exit Mach number to be 1. Thus $L^* = L = 5$ m and $f L^*/D = 2.4$. From the Fanno table, we can get

$$M_{throat} \approx 0.4 \text{ and } \frac{P_{0,throat}}{P_0^*} = 1.59014$$

Therefore,

$$P^* = \frac{P_0^*}{P_{0,throat}} P_{0,throat} \frac{P^*}{P_0^*}$$

$$= \frac{1}{1.59014} (1000) \frac{1}{1.89293} = 332\, kPa$$

where we have used $M = 1$ at the exit. Since P^* is greater than the back pressure, it is clear that the flow at the exit is under-expanded and the Mach number at the exit is indeed equal to one. Proceeding further,

$$\dot{m}(L = 5\,m) = \rho_{throat}\, u_{throat}\, A_{throat}$$

$$= \frac{P_{throat}}{R T_{throat}} M_{throat} \sqrt{\gamma R T_{throat}}\, A_{throat}$$

$$= \frac{P_{throat}}{P_{0,1}} P_{0,1} \sqrt{\frac{\gamma}{R T_0}} \sqrt{\frac{T_0}{T_{throat}}} M_{throat}\, A_{throat}$$

$$= \frac{1}{1.1117} \frac{10^6}{(3.1143 \times 10^{-3})} \sqrt{1.032}\, (0.4)\, (1.9635 \times 10^{-3})$$

$$= \mathbf{2.235\, kg/s}$$

The reduction in mass flow rate is about 37%.

Example 6.5 In an aircraft jet engine fitted with a constant area afterburner and a converging nozzle, post-combustion gas enters the nozzle with a stagnation temperature and pressure of 900 K and 0.5 MPa, when the afterburner is not lit. With the afterburner lit, the stagnation temperature increases to 1900 K with a 15% loss in stagnation pressure at the nozzle inlet. If the mass flow rate has to be maintained at 90 kg/s, determine the required nozzle exit area in both cases. Also, determine the thrust augmentation with afterburner operation, assuming that the engine is on a static test stand at sea level (ambient pressure 0.1 MPa). Assume isentropic process for the nozzle. For the gas, take $\gamma = 4/3$ and $C_p = 1.148$ kJ/kg.K

Solution When the afterburner is not lit, the stagnation pressure at entry to the nozzle is $P_0 = 0.5$ MPa. The critical pressure corresponding to this stagnation pressure is

$$P^* = P_0 \left(\frac{2}{\gamma + 1} \right)^{\frac{\gamma}{\gamma - 1}} = 0.27 MPa$$

Since the ambient pressure $P_\infty = 0.1$ MPa is less than P^*, the nozzle is choked and so the exit static pressure $P_e = P^* = 0.27$ MPa. The mass flow rate is thus given by

$$\dot{m} = \frac{P_0 A_{throat}}{\sqrt{T_0}} \sqrt{\frac{\gamma}{R} \left(\frac{2}{\gamma + 1} \right)^{(\gamma+1)/(\gamma-1)}}$$

When the afterburner is not lit, $P_0 = 0.5$ MPa and $T_0 = 900$ K. Substituting these values, we get

$$A_{throat} = 0.1359 \, m^2$$

Also, the exit static temperature can be evaluated as

$$T_e = T^* = T_0 \left(\frac{2}{\gamma + 1} \right) = 771 \, K$$

and the exit velocity is thus

$$u_e = \sqrt{\gamma R T_e} = 543 \, m/s$$

When the afterburner is lit, $P_0 = 0.5 \times 0.85 = 0.425$ MPa and the critical pressure corresponding to this stagnation pressure is

$$P^* = P_0 \left(\frac{2}{\gamma + 1} \right)^{\frac{\gamma}{\gamma - 1}} = 0.229 MPa$$

Since the ambient pressure $P_\infty = 0.1$ MPa is less than P^*, the nozzle is choked in this case also and so the exit static pressure $P_e = P^* = 0.229$ MPa. With $T_0 = 1900$ K, we can get

$$A_{throat} = 0.1359 \left(\frac{0.5}{0.425} \right) \sqrt{\frac{1900}{900}} = 0.3817 \, m^2$$

The exit static temperature T_e and the exit velocity u_e can be calculated to be 1629 K and 789 m/s in the same manner as before.

From Eq. 6.4, thrust developed is given as

$$\mathcal{T} = \dot{m}\, u_e + (P_e - P_\infty)\, A_e$$

where we have set $u_\infty = 0$, since the engine is on a static thrust stand. Therefore,

Thrust (no afterburner) = 48.87 kN + 23.103 kN = 71.973 kN

and Thrust (with afterburner) = 71 kN + 49.24 kN = 120.24 kN

Thrust augmentation due to afterburner is about 67%. Note that the pressure thrust in both the cases is nonzero due to under-expanded operation of the nozzle.

Example 6.6 In the above example, if a converging diverging nozzle is used instead of a convergent nozzle during normal operation without an afterburner, what would be the thrust?

Solution Assuming that the flow is correctly expanded, exit pressure with the converging diverging nozzle is 0.1 MPa. Thus,

$$\frac{P_0}{P_e} = 5 = \left(1 + \frac{\gamma - 1}{2} M_e^2\right)^{\frac{\gamma}{\gamma - 1}}$$

This can be solved easily to give the exit Mach number $M_e = 1.73$. The exit static temperature T_e can be calculated using

$$T_e = \frac{T_0}{1 + \frac{\gamma - 1}{2} M_e^2} = 602\ K$$

The exit velocity,

$$u_e = M_e \sqrt{\gamma R T_e} = 830\ m/s$$

Therefore, thrust (no afterburner) = 74.7 kN + 0 kN = 74.7 kN. It is clear that the increase in thrust is minimal when we use a convergent divergent nozzle in this case.

Example 6.7 We close this chapter with a very interesting example involving nozzle–nozzle interaction, adapted from White's book (who attributes this problem to a book by Thompson). The schematic is shown in Fig. 6.13. The arrangement consists of two tanks, with the one in the left being larger and two identical converging nozzles. The pressure in the large tank remains constant at 1 MPa and the ambient pressure is 0.1 MPa. We wish to find out whether the nozzles are choked or not under steady-state operation (neglecting heat losses).

Solution To begin with, based on the numerical values given, it is easy to conclude that the static pressure (and the stagnation pressure) in the smaller tank will be well above the critical value of 0.18929 MPa. In fact, it is likely to be close to 1 MPa.

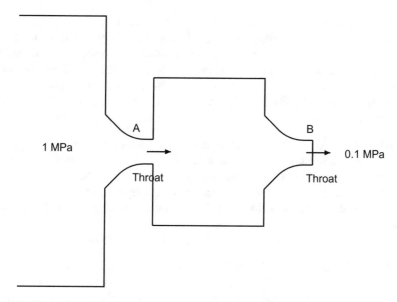

Fig. 6.13 Illustration of Nozzle–Nozzle interaction (Adapted from White)

Consequently, the nozzle on the smaller tank will be choked. Although we can safely assume that the nozzles operate isentropically, the mixing process in the smaller tank introduces some irreversibility in the flow. Due to this irreversibility, the stagnation pressure for the downstream nozzle is less than that for the one located on the larger tank. In addition, T_0, A_{throat} are also the same for both the nozzles. Since, during steady-state operation, the mass flow rates through both the nozzles have to be equal, we can conclude from Eq. 6.10 that the nozzle on the larger tank is not choked.

Note that, during start-up, the nozzle on the larger tank is choked, while the other nozzle is not. As the smaller tank fills up and the pressure in this tank increases, the first nozzle unchokes and the second one chokes (prove this yourself).

Problems

6.1 Consider the two-tank system in Fig. 6.13. Assume, in addition, the stagnation temperature to be 300 K and the throat diameter of the nozzles to be 2.54 cm. Initially, the pressure in the small tank is equal to the ambient pressure. Sketch the variation of the exit pressure, mass flow rate, and the exit Mach number of nozzles A and B with time starting from time 0+ until steady state is reached. Although the profiles can be qualitative, key instants should be marked with numerical values for these quantities. The pressure profiles must be shown together in the same figure using the same axes.

6.2 Consider again the two-tank system in Fig. 6.13. Assume that only nozzle A is present and that it is a convergent divergent nozzle of exit-to-throat area ratio 2 with the same throat diameter as before. Sketch the variation of the exit pressure, mass flow rate, exit Mach number, and the ambient pressure of nozzle A with time starting from time 0+ until steady state is reached. Although the profiles can be qualitative, key instants should be marked with numerical values for these quantities. The pressure profiles must be shown together in the same figure using the same axes.

6.3 Air enters a convergent divergent nozzle of a rocket engine at a stagnation temperature of 3200 K. The nozzle exhausts into an ambient pressure of 100 kPa and the exit-to-throat area ratio is 10. The thrust produced is 1300 kN. Assume the expansion process to be complete and isentropic. Determine (a) the exit velocity and static temperature, (b) mass flow rate, (c) stagnation pressure, (d) throat and exit areas.

   [2207 m/s, 777 K; 589.2 kg/s; 14.2 MPa; 0.0595 m^2, 0.595 m^2]

6.4 A reservoir of volume V initially contains air at pressure P_i and temperature T_i. A hole of cross-sectional area A develops in the reservoir and the air begins to leak out. Develop an expression for the time taken for half of the initial mass of air in the reservoir to escape. Assume that, during the process, the pressure in the reservoir is much higher than the ambient pressure and also that the temperature remains constant.

6.5 Consider a CD nozzle with exit and throat areas of 0.5 m^2 and 0.25 m^2, respectively. The inlet reservoir pressure is 100 kPa and the exit static pressure is 60 kPa. Determine the exit Mach number.

   [0.46]

6.6 Air at a pressure and temperature of 400 kPa and 300 K contained in a large vessel is discharged through an isentropic nozzle into a space at a pressure of 100 kPa. Find the mass flow rate if the nozzle is (a) convergent and (b) convergent divergent with optimum expansion ratio. In both cases, the minimum cross-sectional area of the nozzle may be taken to be 6.5 cm^2.

   [0.6067 kg/s in both cases]

6.7 Air flows in a frictionless, adiabatic duct at M = 0.6 and P_0 = 500 kPa. The cross-sectional area of the duct is 6 cm^2 and the mass flow rate is 0.5 kg/s. If the area of the duct near the exit is reduced so as to form a convergent nozzle, what is the minimum area possible without altering the flow properties in the duct? If the area is reduced to 3/4th of this value, determine the change (if any) in the mass flow rate and the static and stagnation pressure in the duct.

   [5.051 cm^2; 0.125 kg/s, 447.8 kPa and stagnation pressure remains the same]

6.8 A student is trying to design an experimental setup to produce a correctly expanded supersonic stream at a Mach number of 2 issuing into ambient at 100 kPa. For this purpose, the student wishes to use a CD nozzle with the largest possible exit area. There is a 10 m^3 reservoir containing air at 1 MPa and 300 K available in the lab. The nozzle is connected to the reservoir through a settling chamber. The settling chamber is reasonably large and allows the stagnation pressure just ahead of the nozzle to be fixed at the desired value. Determine the largest possible exit area for the nozzle that will allow the student to run the experiment continuously for at least 15 minutes. Neglect frictional losses in the pipes and assume that the temperature of the air remains constant ahead of the nozzle.

[2.587×10^{-5} m^2]

What is the stagnation pressure required to run the nozzle described in the previous question at the desired Mach number?

[782.4 kPa]

What is the stagnation pressure required if the nozzle discharges into a duct (Fig. 6.11a) instead of directly into the ambient? Assume that the duct discharges into the ambient and that there is a normal shock standing at the duct exit. Neglect frictional loss in the duct.

[173.9 kPa]

If a supersonic diffuser is now connected to the end of the duct to diffuse the air to ambient pressure (thereby eliminating the normal shock), what is the stagnation pressure required to drive the flow?

[100 kPa]

6.9 A rocket nozzle produces 1 MN of thrust at sea level (ambient pressure and temperature 100 kPa and 300 K). The stagnation pressure and temperature are 5 MPa and 2800 K. Determine (a) the exit to throat area ratio, (b) exit Mach number, (c) exit velocity, (d) mass flow rate, and (e) exit area. Determine the thrust developed by the nozzle at an altitude of 20 km, where the ambient pressure and temperature are 5.46 kPa and 217 K. Assume the working fluid to be air with $\gamma = 1.4$.

[5.12; 3.2; 1944 m/s; 514.4 kg/s; 0.6975 m^2; 1.066 MN]

6.10 Assume that the rocket nozzle of the previous problem is designed to develop a thrust of 1 MN at an altitude of 10 km. Determine the thrust developed by the nozzle at 20 km, for the same stagnation conditions in the thrust chamber. Take the ambient pressure and temperature at 10 km to be 26.15 kPa and 223 K. Do you expect to see a normal shock in the nozzle divergent portion during sea-level operation?

[1.033 MN]

6.11 An aircraft engine is operating at an ambient pressure and temperature of P_∞ and T_∞. The mass flow rate through the engine is $\dot{m}$ and the air enters with a velocity of u_∞. Consider the following two choices for the nozzle:

• Convergent with the flow choked at the exit, and
• Convergent divergent with the flow at the exit correctly expanded.

In each case, express the thrust in terms of the quantities given above as well the stagnation pressure P_0 and temperature T_0 in the nozzle. Simplify the expressions by substituting $\gamma = 4/3$. Demonstrate that the thrust produced by the convergent divergent nozzle is always greater. You may assume the flow in the nozzle to be isentropic.

6.12 A supersonic diffuser is designed to operate at a freestream Mach number of 1.7. Determine the ratio of mass flow rate through the diffuser when it operates at a freestream Mach number of 2 with a normal shock in front (see Fig. 6.9) to that at the design condition. Assume M=1 at the throat for both cases. Also assume that the freestream **static** conditions remain the same. What is the ratio of the capture area to the throat area in each case?

[1.073; 1.338, 1.218]

6.13 Air enters the combustion chamber of a ramjet engine (Fig. ??) at $T_0 = 1700\,\text{K}$ and $M = 0.3$. How much can the stagnation temperature be increased in the combustion chamber without affecting the inlet conditions? Assume that the combustion chamber has a constant area of cross section.

[3157 K]

6.14 Air enters a constant area combustor followed by a convergent nozzle. Heat addition takes place in the combustor and the flow is isentropic in the nozzle. The inlet Mach number is 0.3, and the throat-to-inlet area ratio is 0.9.

(a) If the stagnation temperature is doubled in the combustor, determine the exit Mach number.
(b) Determine the stagnation temperature rise that will just cause the nozzle to choke.
(c) If the actual rise in stagnation temperature is 10% higher than the value obtained in (b), determine the inlet Mach number.
(d) If the mass flow rate is the same, determine the required change in the inlet stagnation pressure.

[0.58, 2.55, 0.285, 0.953]

6.15 Consider an arrangement consisting of a converging nozzle followed by a smooth, 1 m long pipe. The diameter of the pipe is 0.04 m. The stagnation conditions upstream of the nozzle are 2.5 MPa and 500 K. Determine the mass flow rate if the Mach number at the exit is 1. Assume the flow in the nozzle to be isentropic and that in the pipe to be Fanno flow. Take $f = 0.01$.

[5.11 kg/s]

6.16 A converging diverging frictionless nozzle is connected to a large air reservoir by means of a 20 m long pipe of diameter 0.025 m. The inlet, throat, and exit diameters of the nozzle are, respectively, 0.025 m, 0.0125 m, and 0.025 m. If the air is expanded to the ambient pressure of 100 kPa and $f = 0.02$ in the pipe, determine the stagnation pressure in the reservoir.

[4064 kPa]

In the previous problem, if the nozzle and pipe were interchanged, then determine the stagnation pressure in the reservoir.

[757.76 kPa]

Chapter 7
Oblique Shock Waves

Oblique shock waves are generated in compressible flows whenever a supersonic flow is turned into itself through a finite angle. In some applications such as intakes of supersonic vehicles, ramjet and scramjet engines, the intended objective is to decelerate and compress the incoming air through a series of such oblique shocks thereby eliminating the need for the compressor and the turbine. In other applications, the dynamics of the flow triggers oblique shock waves. This is the case, for instance, when an over-expanded supersonic jet issues from a nozzle into the ambient. Oblique shocks are generated from the corners in the exit plane of the nozzle to compress the jet and increase the static pressure to the ambient value.

Oblique shocks are similar to normal shocks, in that, the flow undergoes a compression process in both cases. However, in the case of the oblique shock wave, unlike a normal shock wave, the flow changes direction after passing through the shock wave. In both cases, the velocity component normal to the shock wave decreases. Since the normal component of velocity is always less than the magnitude of the velocity itself, for a given velocity and temperature of the fluid before the shock, the actual Mach number of the flow as "seen" by the shock wave is less for an oblique shock. Consequently, the stagnation pressure loss across an oblique shock is also lower.

Consider a shock wave moving with speed V_s into quiescent air as shown in the top in Fig. 7.1 (this is the same scenario as shown in Fig. 3.1). If we now switch to a reference frame in which the observer moves with the shock wave at a speed V_s as before, but also *along* the shock wave at a speed V_t, the resulting flow field (after rotation in the counter clockwise direction to make u_1 appear horizontal) is as shown in the bottom of Fig. 7.1. Here, $u_{n,1} = V_s$ and $u_t = V_t$. Subscripts n and t refer to normal and tangential directions respectively to the shock wave. The important point to note is that, now, there is a direction associated with the shock wave. From the velocity triangle, it is clear that the tangential component of the velocity remains the same, but the normal component decreases across the shock wave. As a result, the flow is deflected by an angle θ *towards* the shock wave after passing through it.

© The Author(s) 2021
V. Babu, *Fundamentals of Gas Dynamics*,
https://doi.org/10.1007/978-3-030-60819-4_7

Fig. 7.1 Illustration of an
oblique shock

Observer Stationary

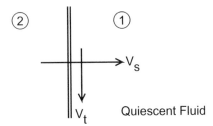

Observer Moving With and Along Shock Wave

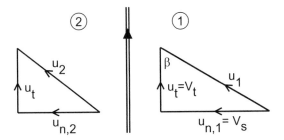

Observer Moving With and Along Shock Wave

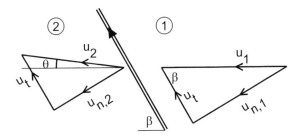

An example of such a flow field is given in Fig. 7.2. Here, a supersonic flow is turned through an angle θ at a sharp corner by means of an oblique shock wave. Note that the shock wave is generated at the corner and propagates into the fluid. Based on the directions of the velocity vectors u_1 and u_2, as well as the direction of the shock wave, it is easy to see that the flow is turned into *itself* in this case. The shock wave illustrated in this figure is called a left running shock wave. The angle β that the shock wave makes with the velocity vector $\mathbf{u_1}$ (**and not necessarily with the horizontal**) is called the shock or wave angle.

Fig. 7.2 Oblique shock
from a sharp corner

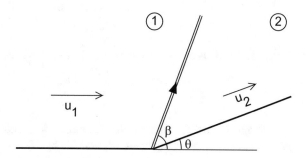

An oblique shock which turns the flow away from itself, would result in the Mach number increasing across the shock wave. As discussed in Chap. 3, such expansion shocks are forbidden by Second Law.

7.1 Governing Equations

The governing equations for this flow (in a frame of reference where the shock is stationary i.e., observer moving with and along the wave) are almost the same as those for normal shock wave. These can be written as

$$\rho_1 u_{n,1} = \rho_2 u_{n,2} \tag{7.1}$$

$$P_1 + \rho_1 u_{n,1}^2 = P_2 + \rho_2 u_{n,2}^2 \tag{7.2}$$

$$h_1 + \frac{1}{2}u_{n,1}^2 + \frac{1}{2}u_t^2 = h_2 + \frac{1}{2}u_{n,2}^2 + \frac{1}{2}u_t^2 \tag{7.3}$$

As before, $T_{0,2} = T_{0,1}$. Comparison of these equations with the governing equations for a normal shock wave shows that u_1 and u_2 from the latter are now replaced by $u_{n,1}$ and $u_{n,2}$.

From the velocity triangles in Fig. 7.1, it is easy to see that

$$u_{n,1} = u_1 \sin \beta, \qquad u_{n,2} = u_2 \sin(\beta - \theta)$$

and

$$u_t = u_1 \cos \beta = u_2 \cos(\beta - \theta)$$

It follows that

$$\frac{u_{n,1}}{u_{n,2}} = \frac{\tan \beta}{\tan(\beta - \theta)}$$

But

$$\frac{u_{n,1}}{u_{n,2}} = \frac{\rho_2}{\rho_1}$$

from Eq. 7.1. This can be rewritten as follows:

$$\frac{u_{n,1}}{u_{n,2}} = \frac{P_2}{P_1}\frac{T_1}{T_2}$$

The right hand side can be written in terms of Mach numbers by using the relations given in Sect. 3.2, after replacing M_1 and M_2 by $M_{n,1}$ and $M_{n,2}$. Thus

$$\frac{u_{n,1}}{u_{n,2}} = \frac{(\gamma+1)M_{n,1}^2}{2+(\gamma-1)M_{n,1}^2}$$

If we equate the two expressions for $u_{n,1}/u_{n,2}$, we get

$$\frac{\tan\beta}{\tan(\beta-\theta)} = \frac{(\gamma+1)M_{n,1}^2}{2+(\gamma-1)M_{n,1}^2}$$

From the velocity triangles in Fig. 7.1,

$$M_{n,1} = M_1\sin\beta, \qquad M_{n,2} = M_2\sin(\beta-\theta)$$

Upon substituting this into the above relationship, we get

$$\frac{\tan\beta}{\tan(\beta-\theta)} = \frac{(\gamma+1)M_1^2\sin^2\beta}{2+(\gamma-1)M_1^2\sin^2\beta}$$

With a little of algebra, this can be written as

$$\tan\theta = 2\cot\beta\left(\frac{M_1^2\sin^2\beta-1}{M_1^2(\gamma+\cos 2\beta)+2}\right) \tag{7.4}$$

This relation is known as the $\theta-\beta-M$ relation. Given any two of the three quantities, M_1, θ and β, this relation can be used to determine the third quantity.

7.2 θ-β-M Curve

Solutions of the $\theta-\beta-M$ equation for different values of the upstream Mach number M_1 are illustrated in Fig. 7.3. This figure is meant to be an illustration for elucidating the important features of the $\theta-\beta-M$ relation. For calculation purposes, a more accurate representation in graphical or tabular form is usually available (see Table A.5 given towards the end). Several important features that emerge from this illustration are:

- For any upstream Mach number M_1, there exists a maximum value of deflection angle $\theta = \theta_{max}$ (indicated in Fig. 7.3 by filled circles). If the required flow deflection angle is higher than this value, then the flow turning cannot be accomplished by means of an attached oblique shock.
- For a given value of upstream Mach number M and flow deflection angle θ, there are two possible solutions—a weak and a strong shock wave solution. The two solutions are separated by the θ_{max} point. The line separating the weak and strong shock solutions is indicated in Fig. 7.3 by a dashed line. The wave angle for the weak shock solution is less than that of the strong shock. In reality, attached strong shocks are seldom, if ever, seen. Almost all of the attached oblique shocks seen in real life applications are weak shocks. However, detached shocks can be partly strong and partly weak as described in the next section. It is of interest to note that, for a given value of M_1, the value of β corresponding to $\theta = \theta_{max}$ is given as (see Hodge & Koenig)

$$\sin^2 \beta_{\theta=\theta_{max}} = \frac{1}{\gamma M_1^2} \left[\frac{\gamma+1}{4} M_1^2 - 1 \right.$$

$$\left. + \sqrt{(\gamma+1)\left(1 + \frac{\gamma-1}{2} M_1^2 + \frac{\gamma+1}{16} M_1^4\right)} \right]$$

Using this value for β, the corresponding value for θ_{max} can be obtained from Eq. 7.4.

- For the strong shock solution, M_2 is always less than one. For the weak shock, however, M_2 is almost always greater than one, except when the flow deflection angle is close to θ_{max}. This region, where M_2 is less than one for the weak shock, lies between the dashed and the dot-dashed line in Fig. 7.3. Note that even when M_2 is greater than one, $M_{n,2}$ is less than one, since $M_{n,1}$ and $M_{n,2}$ are related through the normal shock relationship.
- The strong shock portion of all the $M = constant$ curves intersect the abscissa at $\beta = 90°$. This corresponds to a deflection angle of $0°$ with the shock wave normal to the flow direction, which is nothing but the normal shock solution.
- The weak shock portion of the $M = constant$ curves, on the other hand, intersect the abscissa at different points. The point of intersection represents an infinitesimally weak shock wave (called a Mach wave) which deflects the flow through an infinitesimally small angle. The compression process is isentropic in this case, and this type of compression wave is the subject matter of the next chapter. The wave angle corresponding to the point of intersection is called the Mach angle, μ, and is equal to $\sin^{-1}(1/M)$. The significance of this angle is discussed in the next chapter.
- Each $M = constant$ curve wherein $\mu \le \beta \le 90°$ in Fig. 7.3, depicts *all possible* compressive wave solutions, namely, Mach wave $\rightarrow$ Weak Oblique shock wave $\rightarrow$ Strong Oblique shock wave $\rightarrow$ Normal shock wave in sequence.

Fig. 7.3 Illustration of the
$\theta - \beta - M$ curve

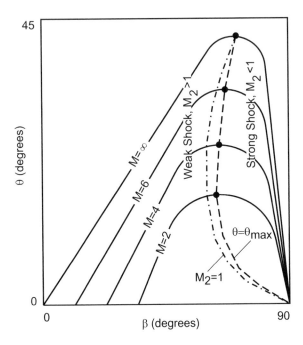

Based on the above observations, we can make some inferences on the loss of stagnation pressure and entropy change across a shock wave. From Eq. 2.17, we can
write

$$\frac{P_{0,2}}{P_{0,1}} = \frac{P_{0,2}}{P_2} \frac{P_2}{P_1} \frac{P_1}{P_{0,1}} = \left(\frac{1 + \frac{\gamma - 1}{2} M_2^2}{1 + \frac{\gamma - 1}{2} M_1^2} \right)^{\frac{\gamma}{\gamma - 1}} \frac{P_2}{P_1}$$

From this expression, it is easy to see that, for a given value of M_1 and P_1, loss of
stagnation pressure ($P_{0,1} - P_{0,2}$) increases with decreasing M_2 and increasing P_2.
Note that, since M_2 and P_2 are related, a decrease in the former automatically results
in an increase in the latter. With this in mind, if we follow a $M = constant$ curve in
Fig. 7.3, we can see that since M_2 decreases along this curve, the loss of stagnation
pressure increases from zero for $\beta = \mu$ to a maximum for $\beta = 90°$. It is easy to see
from Eq. 2.19, that entropy change follows the same trend.

7.3 Illustration of the Weak Oblique Shock Solution on a T-s Diagram

Although Fig. 7.3 is very informative, it is not possible to depict thermodynamic states in this figure. Since oblique shock waves are encountered extensively, it would be worthwhile depicting the corresponding thermodynamic states on a T-s diagram. Of course, the strong oblique shock solution is almost identical to the normal shock solution illustrated in Fig. 3.3 from a thermodynamic perspective and hence the T-s diagram for the strong oblique shock solution will not be presented here. However, the weak oblique shock wave solution differs substantially and we turn to the illustration of the same on a T-s diagram.

Based on the discussion in the previous section, for a weak oblique shock wave, $M_1 > M_2 > 1$, $M_{n,1} > 1 > M_{n,2}$, $P_2 > P_1$, $T_2 > T_1$, $T_{0,2} = T_{0,1} = T_0$ and $P_{0,2} < P_{0,1}$. Note that from Eq. 7.3,

$$T_0 = T + \frac{1}{2}u_n^2 + \frac{1}{2}u_t^2 = T + \frac{1}{2}u^2$$

where the last equality follows from the velocity triangles shown in Fig. 7.1.

States 1 and 2 for a weak oblique shock wave are illustrated in Fig. 7.4. Since the static temperature of state 1, $T_1 < T^*$, where $T^* = 2T_0/(\gamma + 1)$, state 1 is supersonic. For the same reason, state 2 is also seen to be supersonic in this figure.

Equation 7.3 may be simplified to obtain

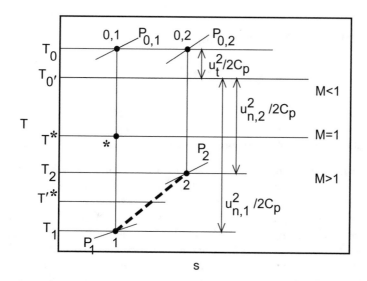

Fig. 7.4 Illustration of the weak oblique shock solution on a T-s diagram

$$T_{0'} = T_1 + \frac{1}{2}u_{n,1}^2 = T_2 + \frac{1}{2}u_{n,2}^2$$

where only the normal components of the velocity appear. The isotherm corresponding to $T_{0'}$ is also shown in Fig. 7.4. The sonic temperature calculated using $T_{0'}$ is given as $T'^* = 2T_{0'}/(\gamma + 1)$. It can be seen that, since $T_1 < T'^*$, $M_{n,1} > 1$ and since $T_2 > T'^*$, $M_{n,2} < 1$.

Example 7.1 Supersonic flow at $M = 3$, $P = 100$ kPa and $T = 300$ K is deflected through $20°$ at a compression corner. Determine the shock wave angle and the flow properties downstream of the shock.

Solution For $M_1 = 3$, from the isentropic table,

$$\frac{P_{0,1}}{P_1} = 36.73 \text{ and } \frac{T_{0,1}}{T_1} = 2.8$$

Thus, $P_{0,1} = 3.673$ MPa and $T_{0,1} = 840$ K.

For $M_1 = 3$ and $\theta = 20°$, from the oblique shock table, we get $\beta = \mathbf{37.76°}$. Therefore, $M_{n,1} = M_1 \sin\beta = 1.837$.

From the normal shock table, for $M_{n,1} = 1.837$, we get

$$\frac{P_2}{P_1} = 3.77036, \quad \frac{T_2}{T_1} = 1.56702 \text{ and } M_{n,2} = 0.608$$

Hence,

$$\mathbf{M_2} = \frac{M_{n,2}}{\sin(\beta - \theta)} = \mathbf{2.0}$$

$\mathbf{P_2 = 377}$ **kPa and** $\mathbf{T_2 = 470}$ **K**. Also,

$$\mathbf{P_{0,2}} = \frac{P_{0,2}}{P_2}P_2 = (7.82445)(377) = \mathbf{2.95\,MPa}$$

and $\mathbf{T_{0,2}} = T_{0,1} = \mathbf{840\,K}$.

The loss of stagnation pressure for this case is about 20 percent. Had this been a normal shock, the loss of stagnation pressure would have been 67%.

Example 7.2 Mixed compression supersonic intakes (Fig. 7.5) are widely used in supersonic vehicles, ramjet and scramjet engines owing to their superior off-design performance when compared with internal compression intakes. Here, external compression is achieved by means of oblique shocks generated from properly designed ramps and terminated by a normal shock. This is followed by internal compression. The intake shown in Fig. 7.5 is designed for operation at $M_\infty = 3$, $P_\infty = 15$ kPa and $T_\infty = 135$ K. The ramp angles are $15°$ and $30°$ respectively. For the critical mode of operation, determine the mass flow rate through the intake, cross-sectional area at the beginning of internal compression and the total pressure recovery.

Fig. 7.5 2D mixed compression supersonic intake

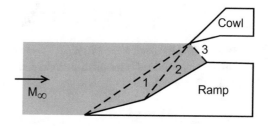

Solution Critical mode of operation refers to the situation depicted in Fig. 7.5, when the oblique shocks intersect at the cowl leading edge. The cross-sectional view of the streamtube that enters the intake is shown in this figure as a shaded region. Based on the dimensions given in the figure, the capture area (defined as the freestream cross-sectional area of the streamtube that enters the intake) can be evaluated as $0.0375 \times 1\, m^2$. Thus,

$$\dot{m} = \rho_\infty\, u_\infty\, A_\infty$$

$$= \sqrt{\frac{\gamma}{RT_\infty}}\, P_\infty\, M_\infty\, A_\infty$$

$$= \mathbf{10.11\, kg/s}$$

From the oblique shock table, for $M_\infty = 3$ and $\theta_\infty = 15°$, we get $\beta_\infty = 32.32°$. Therefore, $M_{n,\infty} = M_\infty \sin\beta_\infty = 1.6$. From the normal shock table, for $M_{n,\infty} = 1.6$, we get

$$\frac{P_1}{P_\infty} = 2.82\,, \quad \frac{T_1}{T_\infty} = 1.38797 \quad \text{and} \quad M_{n,1} = 0.668437$$

Hence,

$$M_1 = \frac{M_{n,1}}{\sin(\beta_\infty - \theta_\infty)} = 2.245$$

$P_1 = 42.3$ kPa and $T_1 = 187$ K.

From the oblique shock table for $M_1 = 2.245$ and $\theta_1 = 30° - 15° = 15°$, we get $\beta_1 \approx 40°$.

Therefore, $M_{n,1} = M_1 \sin\beta_1 = 1.443$. From the normal shock table, for $M_{n,1} = 1.443$, we get

$$\frac{P_2}{P_1} = 2.25253\,, \quad \frac{T_2}{T_1} = 1.28066 \quad \text{and} \quad M_{n,2} = 0.723451$$

Therefore,

$$M_2 = \frac{M_{n,2}}{\sin(\beta_1 - \theta_1)} = 1.712$$

$P_2 = 95.282$ kPa and $T_2 = 240$ K.

This is followed by a terminal normal shock. From the normal shock table, for $M_2 = 1.712$, we get

$$\frac{P_3}{P_2} = 3.2449, \quad \frac{T_3}{T_2} = 1.465535, \quad M_3 = 0.638 \text{ and } \frac{P_{0,3}}{P_{0,2}} = 0.8515385$$

Therefore, $P_3 = 309.18$ kPa and $T_3 = 352$ K.

Total pressure recovery at the beginning of internal compression is

$$\frac{P_{0,3}}{P_{0,\infty}} = \frac{P_{0,3}}{P_{0,2}} \frac{P_{0,2}}{P_2} \frac{P_2}{P_1} \frac{P_1}{P_\infty} \frac{P_\infty}{P_{0,\infty}}$$

$$= \frac{(0.8515385)(5.049225)(2.25253)(2.82)}{(36.7327)} = 0.74$$

At the beginning of internal compression, the mass flow rate is

$$\dot{m} = \rho_3 \, u_3 \, A_3$$

$$= \sqrt{\frac{\gamma}{RT_3}} \, P_3 \, M_3 \, A_3$$

$$\Rightarrow A_3 = \mathbf{0.0138 \, m^2}$$

Example 7.3 A converging diverging nozzle with an exit to throat area ratio of 2.637 operates in an over-expanded mode and exhausts into an ambient pressure of 100 kPa (see Fig. 7.6). The inlet stagnation conditions are 300 K and 854.5 kPa. Determine the flow properties at states 2, 3 and 4. Also, find out the angle made by the edge of the jet with the horizontal. (Adapted from Hodge and Koenig).

Solution Given $A_2/A^* = 2.637$, $P_{0,1} = P_{0,2} = 854.5$ kPa and $T_0 = 300$K.
From the isentropic table, for the given area ratio, we get

$$\mathbf{M_2 = 2.5}, \quad \frac{P_{0,2}}{P_2} = 17.09 \text{ and } \frac{T_0}{T_2} = 2.25$$

Therefore, $\mathbf{P_2 = 50 \, kPa}$ and $\mathbf{T_2 = 133 \, K}$.

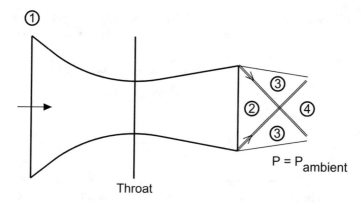

Fig. 7.6 Illustration for worked example

Since the jet is over-expanded, it undergoes compression outside the nozzle by the oblique shocks generated from the trailing edge of the nozzle. Static pressure in region 3 is the same as the ambient pressure. Therefore, **$P_3 = 100$ kPa**.

From the normal shock table, for $P_3/P_2 = 2$, we get

$$M_{n,2} = 1.36, \quad M_{n,3} = 0.7572 \quad \text{and} \quad \frac{T_3}{T_2} = 1.229$$

Since $M_{n,2} = M_2 \sin \beta_2$, we can get $\beta_2 = 33°$.

Also, **$T_3 = 164$ K** from above. Since, $T_{0,3} = 300$ K, we get $T_{0,3}/T_3 = 1.8308$. From the isentropic table, for this value of $T_{0,3}/T_3$, we can get **$M_3 \approx 2.04$**. It follows that

$$\mathbf{P_{0,3}} = \frac{P_{0,3}}{P_3}\,P_3 = (8.32731)(100) = \mathbf{832.731\ kPa}$$

From

$$M_3 = \frac{M_{n,3}}{\sin(\beta_2 - \theta_2)}$$

we get $\theta_2 = \mathbf{11.2118°}$. The edge of the jet makes an angle of 11.2118° (clockwise) with the horizontal.

Since the flow is deflected through 11.2118° in region 3, the shock wave angle for calculation of properties in region 4 is $\beta_3 = 33° + 11.2118° = 44.2118°$. Thus, with respect to this wave, $M_{n,3} = M_3 \sin \beta_3 = 1.4225$.

From the normal shock table, for $M_{n,3} = 1.4225$, we get

$$M_{n,4} = 0.729425, \quad \frac{P_4}{P_3} = 2.2024 \quad \text{and} \quad \frac{T_4}{T_3} = 1.27085$$

Hence, $\mathbf{P_4 = 220.24\,kPa}$ and $\mathbf{T_4 = 208\,K}$. Since $T_{0,4} = 300\,K$, it follows that $T_{0,4}/T_4 = 1.441$, and from the isentropic table, we get $\mathbf{M_4 \approx 1.48}$. It follows that

$$\mathbf{P_{0,4}} = \frac{P_{0,4}}{P_4}\, P_4 = (3.56649)(220.24) = \mathbf{785.48\,kPa}$$

From

$$M_4 = \frac{M_{n,4}}{\sin(\beta_3 - \theta_3)}$$

we get $\theta_3 = \mathbf{14.6834°}$. Since the pressure in region 4 is higher than atmospheric, the flow expands further downstream. This sequence of alternate expansion and compressions goes on for a few jet diameters, producing the characteristic "shock diamond" pattern in the jet.

7.4   Detached Shocks

So far, we have looked at examples involving attached oblique shocks. In real life applications in aerodynamics such as flow around blunt bodies, this is no longer possible and the shock becomes detached. It is of interest to look at the structure of such detached shocks to gain a better understanding of the flow field.

Figure 7.7 illustrates the flow around a wedge[1] with freestream Mach number M_1. When the wedge half-angle θ is less than θ_{max}, the oblique shocks on the upper and lower surface remain attached to the tip of the wedge. When θ becomes greater than θ_{max}, the shocks detach and a curved bow shock stands in front of the wedge. The stand-off distance depends upon M_1 and θ. The flow deflection angle across the bow shock is zero along the centerline and so the change in properties across the shock wave at this point corresponds to that across a normal shock. Consequently, the shock is also the strongest at this location[2]. Far away from the centerline, where the free stream does not feel the presence of the wedge, the bow shock becomes infinitesimally weak and the flow deflection angle also becomes infinitesimally small. It thus becomes clear that the strength of the shock decreases as one moves outwards along the shock from the centerline. Furthermore, flow deflection angle which is zero at the centerline increases, reaches a maximum and then again decreases to zero upon reaching the undisturbed free stream. Based upon these considerations, the structure of the bow shock can be surmised to be as follows: Normal shock →

---

[1] It is possible to look at the cross-sectional view in this figure and assume that it might also represent flow around a cone. The resemblance is only superficial, for the flow around a wedge is two-dimensional, whereas, flow around a cone is inherently three-dimensional.

[2] Increase in temperature due to shock heating is also quite high in this region. In some cases such as the space shuttle when it re-enters the atmosphere, for instance, the temperature rise is high enough to ionize the air. Reentry vehicles thus require special thermal protection such as ceramic or ablative coatings in the nose region to withstand such high temperatures.

Fig. 7.7 Illustration of attached ($\theta < \theta_{max}$) and detached ($\theta > \theta_{max}$) oblique shock

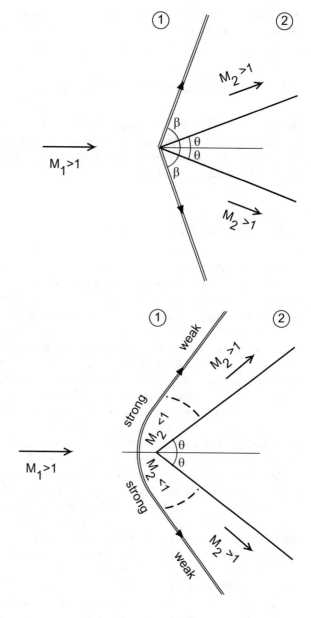

Strong oblique shock $\rightarrow$ Weak oblique shock. This is illustrated in Fig. 7.7. The sonic line which separates the strong and weak shock portion is shown in this figure as a dashed line. Interestingly enough, the same structure can be seen in Fig. 7.3 if we follow the $M_1 = constant$ curve, from $\beta = 90°$ to $\beta = \mu$. It is interesting since the solution shown in Fig. 7.3, is for attached shocks. In the case of a bow shock, the property changes and the flow field has to be calculated by means of computational techniques, but the similarity to the attached shock solution allows us to make quick estimates of these quantities.

7.5   Reflected Shocks

Whenever an oblique shock impinges on a boundary, it is reflected from the boundary in some manner. This boundary can either be a wall or a jet boundary as in Fig. 7.6. We consider the former case here and defer the latter case to the next chapter, as it requires knowledge of expansion fans.

7.5.1   Reflection from a Wall

Reflection of an oblique shock wave is illustrated in Fig. 7.8. Here, an oblique shock is generated from a concave corner on the bottom wall and impinges on the top wall. The flow is deflected upward by an angle θ (i.e., towards the shock wave). If we now consider the flow in a region very close to the top wall, it is clear that the flow here has to be parallel to the wall. Thus, the reflected wave (whatever its nature maybe) has to deflect the flow *downward* through the same angle θ. Since it is a reflected wave, the direction of the wave is now from top to bottom, and so the required flow deflection is towards the reflected wave. Hence the reflected wave is also an oblique shock. Thus, an oblique shock is reflected from a wall as an oblique shock. This type of reflection is known as a regular reflection and is shown in the top in Fig. 7.8. Note that $M_1 > M_2 > M_3$ and so, from Fig. 7.3, $\beta_2 > \beta_1$.

Since the Mach number decreases across the incident shock wave, regular reflection is possible only if the required flow deflection θ is less than θ_{max} corresponding to M_2. From Fig. 7.3, it can be seen that θ_{max} decreases with decreasing Mach number. If, in a particular situation the required θ is greater than θ_{max}, then we do not have a regular reflection but the so-called Mach reflection. This is illustrated in the bottom in Fig. 7.8. The considerations on the velocity vector close to the top wall still remain the same as before. But, now, the incident oblique shock does not extend right up to the top wall but leads to a normal shock being generated near the top wall. The flow after passing through the normal shock remains parallel to the wall. A reflected oblique shock and a slip line[3] is generated from the point of intersection of the incident oblique shock wave and the normal shock wave. The flow field in the downstream region is more complicated than what is shown in this figure and has to be calculated numerically. In real flows, where viscous effects are present, there is a boundary layer near the wall. Impingement and reflection of an oblique shock in

---

[3]A slip line is a discontinuity in the flow field across which some properties such as velocity and entropy are discontinuous, whereas the pressure is continuous. It is thus similar to a shock wave in the former aspect but is different in the latter aspect.

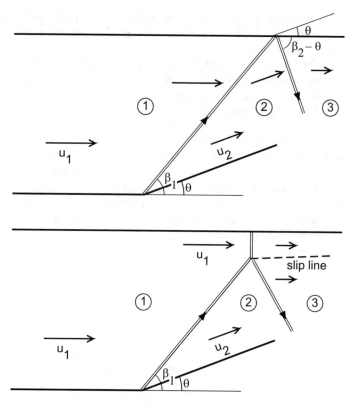

Fig. 7.8 Illustration of reflection of an oblique shock from a wall. Regular reflection (top) and Mach reflection (bottom)

such cases result in the so-called shock boundary layer interaction and the flow field is much more complicated than the scenario that has been outlined above.

Problems

7.1 For the geometry shown in Worked Example 7.2, determine the mass flow rate through the intake and the total pressure recovery for the sub-critical mode of operation when $M_\infty = 2.5$. Compare these values with those of the critical mode. Sketc.h the stream tube that enters the intake in the sub-critical mode and prove that the mass flow rate has to be lower based on geometric considerations.

[6.84 kg/s, 0.893]

7.2 Air is flowing at $M = 2.8$, 100 kPa and 300 K through a frictionless adiabatic duct shown in Fig. 7.8. If the flow turning angle is 15°, determine the static and stagnation properties at sect. 7.3.

If the initial Mach number were 1.5, do you expect a regular reflection?

For a duct inlet Mach number of $M = 2.8$, what is the largest value of the turning angle θ that will give at least three regular reflections?

[601 kPa, 523 K, 2346.1 kPa, 770.4 K; No; 14°]

7.3 A 2D supersonic inlet (Fig. 7.9) is constructed with two ramps each of which deflects the flow through 15°. Following the second oblique shock, a fixed throat inlet is used for internal compression. The inlet is designed to start for a flight Mach number of 2.5. Determine the stagnation pressure recovery, assuming (a) that the inlet starts (no normal shock at the entrance) and (b) that the inlet does not start (normal shock at the entrance). Also, determine the ratio of the entrance area to throat area. Assume that the throat Mach number is 1 in both the cases.

[0.893; 0.87; 1.094, 1.06]

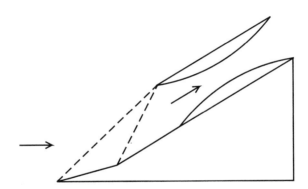

Fig. 7.9 Illustration of a 2D mixed inlet with a fixed throat

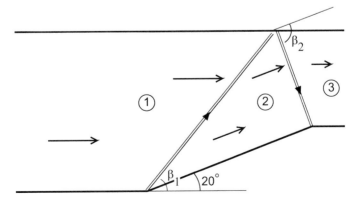

Fig. 7.10 Supersonic flow through a duct with a compression corner

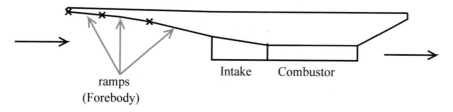

ramps
(Forebody)

Fig. 7.11 Illustration of a scramjet engine

7.4 One way of reducing multiple reflected shocks in the duct shown in Fig. 7.8 is to orient the downstream corner in the lower wall so that the reflected wave is exactly terminated there (see Fig. 7.10). If air enters the duct in Fig. 7.10 at $M = 3$, 100 kPa and 300 K and the height of the duct at sect. 7.1 is 1m, determine the required height at sect. 7.3 and also the static and stagnation properties at sect. 7.3.

[0.342 m, 1066 kPa, 651 K, 2654 kPa, 840 K]

7.5 Consider the forebody, intake and the combustor for a conceptual scramjet engine shown in Fig. 7.11. The freestream conditions are $M = 5$, $P_\infty = 830\,\text{N/m}^2$ and $T_\infty = 230\,\text{K}$. The ramp angles are 5°, 10° and 15° from the horizontal. The engine uses kerosene fuel (calorific value = 45 MJ/kg). The air-fuel ratio on a mass basis is 80. The cross-section at the intake entrance is 920 mm x 250 mm and at combustor entrance 920 mm x 85 mm. Determine (a) the mass flow rate through the engine and (b) the Mach number, stagnation pressure recovery at the combustor inlet and outlet. Sketc.h the process undergone by the fluid on a T-s diagram indicating all the static and stagnation states clearly. Assume critical mode of operation. Ignore expansion of the flow around the corner at the entry to the combustor.

[13.18 kg/s; 2.56, 0.952; 1.18, 0.41]

Chapter 8
Prandtl-Meyer Flow

In the previous chapters, we looked at flow across normal and oblique shock waves. In both cases, the fluid undergoes compression and the flow decelerates. There is also an attendant loss of stagnation pressure. It was also shown that an "expansion shock" solution, while mathematically possible, is forbidden by the second law of thermodynamics as the entropy decreases across such a shock wave, which is adiabatic.

In contrast, for the flow across an acoustic wave, that we looked at in Sect. 2.2, all the changes in properties occur isentropically. Furthermore, there is no restriction on the nature of the process—it can be a compression or an expansion process. Of course, we must keep in mind that the wave that was considered in Sect. 2.2, moves at the speed of sound. In this chapter, we explore whether it is possible to achieve changes in the properties of a supersonic flow, in a similar manner.

8.1 Propagation of Sound Waves and the Mach Wave

Consider a point source of disturbance moving in a compressible medium as shown in Fig. 8.1. The point source is shown in this figure as a filled circle and it moves from right to left. As it moves, the point source generates acoustic (sound) waves which travel in spherical fronts[1] at a speed equal to a, the speed of sound in the medium under consideration. The relative positions of the point source and the wave fronts at different instants of time are illustrated in Fig. 8.1.

Let us imagine a stationary observer located on the ground in front of the source. This observer will notice that, when the source moves at a speed less than the speed of sound, then the sound waves arrive before the source passes overhead. When the

---

[1]This is true only in 3D. In 2D, the front is cylindrical rather than spherical with the axis of the cylinder perpendicular to the page.

© The Author(s) 2021
V. Babu, *Fundamentals of Gas Dynamics*,
https://doi.org/10.1007/978-3-030-60819-4_8

Fig. 8.1 Generation and
propagation of sound waves
in a compressible medium
due to a moving point source.
Source speed subsonic (top)
and sonic (bottom). Source
speed supersonic

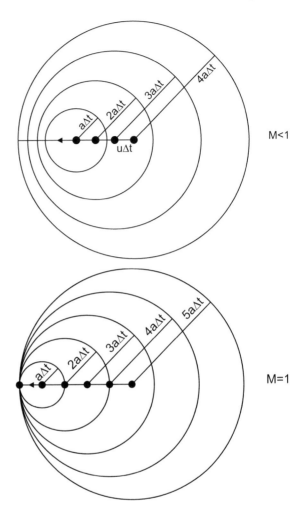

source moves with speed equal to the speed of sound, then the observer will notice
that the sound waves arrive at the same instant when the source passes overhead.
When the source speed is greater than the speed of sound, then the observer will
notice that the source passes overhead before the sound waves arrive. In this case,
any observer, initially situated outside the cone shaped region (wedge shaped region
in the case of 2D) with the source at its apex will first notice the source passing
overhead. Sound waves arrive at the observer location only when the cone shaped
region moves to enclose the observer. It is important to note that the observer is
stationary and the cone shaped region moves along with the source. Hence it is clear
that information about the source is confined to this region and thus it is called the
"zone of action". Similarly, the source itself is aware only of events that happen

Fig. 8.1 (continued)

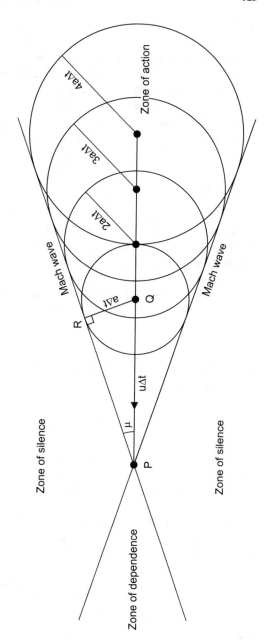

within a cone shaped region called "zone of dependence", which is a mirror image of the zone of action. The remaining region is called the "zone of silence".

Across the surface of the cone itself, velocity, density, pressure and other properties of the fluid change by an infinitesimal amount but in a discontinuous manner (similar to the situation considered in Sect. 2.2). Thus the surface of the cone can be visualized as a wave front and it is usually called the Mach wave. From the geometric construction in Fig. 8.1, it is clear that the surface of the cone is tangential to the wave fronts, and so the semi-vertex angle of the cone, from $\triangle PQR$ is,

$$\mu = \sin^{-1}\left(\frac{a\Delta t}{u\Delta t}\right) = \sin^{-1}\left(\frac{1}{M}\right) \tag{8.1}$$

This angle is called the Mach angle. Also, since the Mach wave is tangential to the acoustic wave front, the normal velocity component of the fluid approaching the Mach wave is equal to a. If we now switch to a reference frame in which the source is stationary, then the flow approaches the source of disturbance at a supersonic speed. A Mach cone is generated with the source at the apex, and semi-vertex angle equal to the Mach angle, μ.[2] This frame of reference is extremely useful for describing supersonic flow around corners, as will be seen next.

8.2 Prandtl Meyer Flow Around Concave and Convex Corners

Consider supersonic flow over a smooth concave corner as shown in Fig. 8.2. Let us assume that the curved portion of the surface is composed of an infinite number of small segments. We can now visualize each segment to be a point source of disturbance similar to the one discussed in the previous section. Thus, each segment generates a Mach wave in the direction shown in the figure (only four such waves are shown in the figure for the sake of clarity). Since the flow in this case is turned *towards* the wave, the velocity magnitude decreases across each wave (see the velocity triangle in Fig. 8.4) and so the corner is a compressive corner. However, the compression process is isentropic, since the velocity magnitude decreases only infinitesimally across each Mach wave. As the Mach number decreases continuously, the Mach angle μ increases continuously as shown in the figure. Consequently, all the Mach waves converge and intersect at a point further away from the surface and coalesce to form an oblique shock. A slip line is generated at the point of coalescence and this separates the flow that has passed through the oblique shock (and so has a higher entropy) and the flow near the surface that has been compressed isentropically. The actual flow in the vicinity of this point of coalescence is quite complex and outside the scope of this book.

---

[2]It is important to understand that a Mach wave is generated only for a point disturbance. For a finite sized disturbance, an oblique shock wave will be generated.

Fig. 8.2 Supersonic flow
around a smooth concave
corner

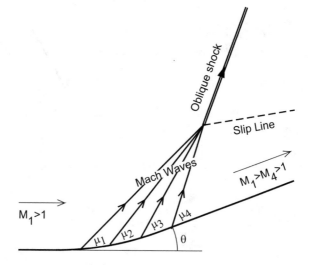

Supersonic flow around a smooth convex corner is illustrated in Fig. 8.3. By using
the same analogy as before, it can be seen that the flow turning is accomplished
through Mach waves generated from the corner, each one turning the flow by an
infinitesimal amount. In this case, the flow is turned *away* from the waves and so the
magnitude of the velocity increases across each wave (see the velocity triangle in
Fig. 8.4). Thus, flow turning around a convex corner is an expansion process. Since
the Mach number increases progressively, the Mach angle decreases as shown in
Fig. 8.3 and the Mach waves diverge from each other in contrast to the previous case.
Hence, the Mach waves cannot coalesce and form an "expansion shock".

A special case of the expansion corner is the sharp convex corner shown in
Fig. 8.3. An expansion fan centered at the corner is generated in this and the
flow expands as it goes through the fan. The expansion process is again isentropic. It
is important to note that there is no analogous situation in the case of a compression
corner. That is, compression (and hence flow turning) in a sharp concave corner is
always accomplished through an oblique shock (Fig. 7.2). This can be seen from
Fig. 8.2 also. As the corner becomes sharper, the oblique shock, which was located
away from the surface, moves close to the surface and eventually stands at the corner
itself.

8.3   Prandtl Meyer Solution

The salient features of the flow across a Mach wave are illustrated in Fig. 8.4 for
both a compression and an expansion wave. Imagine an observer sitting at point R in
Fig. 8.1. For this observer, the freestream appears to approach with a velocity u_1
(speed of the point disturbance in the laboratory frame of reference). At the same time,

Fig. 8.3 Supersonic flow
around a smooth convex
corner (top) and sharp
convex corner (bottom)

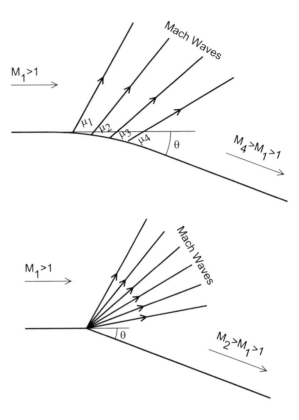

this observer is also moving along the direction QR in Fig. 8.1 (normal to the Mach cone) with a speed equal to a_1 (as measured in the laboratory frame of reference). Hence, the observer perceives the approaching flow to have two components—one along the freestream direction and one normal to the surface of the Mach cone. This is depicted in the velocity triangle ahead of the Mach wave in Fig. 8.4. Since u_t remains the same across the Mach wave, the velocity triangles before and after the wave can be combined. The combined velocity triangles are shown enlarged in Fig. 8.4. Note that the flow deflection angle dv as well as the change in velocity and other properties are infinitesimally small. Thus, we have written $u_2 = u_1 \pm du$, and $u_{n,2} = u_{n,1} \pm du_n$. Also, $u_{n,1} = a_1$, as discussed earlier. From $\triangle OPQ$, $PQ = u_1 \sin dv \approx u_1 dv$. Note that $\angle RPQ = \mu_1 \pm dv$. Hence, from $\triangle PQR$, we get $PQ = du_n \cos(\mu_1 \pm dv) \approx du_n \cos \mu_1$. Upon equating the two expressions for PQ, we get

$$dv = \frac{du_n}{u_1} \cos \mu_1$$

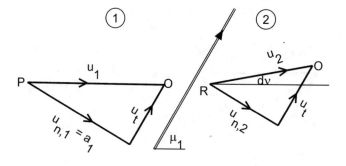

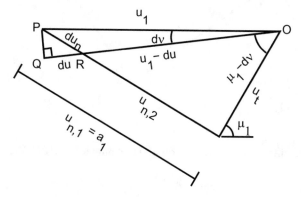

Fig. 8.4 Illustration of a compression wave and combined velocity diagram. Expansion wave and combined velocity diagram

Also, from $\triangle PQR$, $QR = du = du_n \sin(\mu_1 \pm dv) \approx du_n \sin \mu_1$. Therefore,

$$du_n = \frac{du}{\sin \mu_1}$$

If we substitute this into the above expression for dv, we get

$$dv = \frac{du}{u_1} \cot \mu_1 = \frac{du}{u} \sqrt{M_1^2 - 1} \qquad (8.2)$$

where we have used the fact that $\sin \mu_1 = 1/M_1$. Let us now express du/u_1 in terms of M_1. The stagnation temperature is given as,

$$T_0 = T_1 + \frac{u_1^2}{2C_p}$$

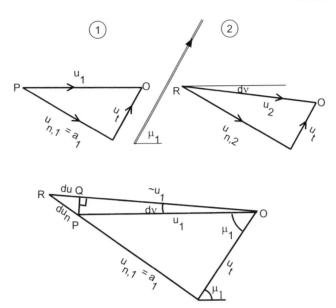

Fig. 8.4 (continued)

If we take the differential of this expression (keeping in mind that T_0 is a constant), we get

$$dT + \frac{\gamma - 1}{\gamma R} u_1 du = 0$$

where the subscripts have been dropped from the differentials. This can be written as,

$$\frac{dT}{T_1} = -(\gamma - 1)M_1^2 \frac{du}{u_1}$$

where we have used $M_1 = u_1/\sqrt{\gamma R T_1}$. Further, if we take the differential of this expression for the Mach number, we get

$$\frac{dM}{M_1} = \frac{du}{u_1} - \frac{dT}{T_1}$$

If we combine the last two expressions, we get

$$\frac{dM}{M_1} = 2\left(1 + \frac{\gamma - 1}{2}M_1^2\right)\frac{du}{u_1}$$

This can be simplified to yield

$$\frac{du}{u_1} = \frac{dM_1^2}{2M_1^2 \left(1 + \frac{\gamma - 1}{2}M_1^2\right)}$$

Equation (8.1) can thus be written as

$$dv = \frac{\sqrt{M_1^2 - 1} \, dM_1^2}{2M_1^2 \left(1 + \frac{\gamma - 1}{2}M_1^2\right)}$$

Note that $dv = 0$, when $M_1 = 1$, which is consistent with what we discussed in Sect. 2.2. If we now integrate this equation from $M_1 = 1$ to any M, we get

$$v = \sqrt{\frac{\gamma + 1}{\gamma - 1}} \tan^{-1} \sqrt{\frac{\gamma - 1}{\gamma + 1}(M^2 - 1)} - \tan^{-1} \sqrt{M^2 - 1} \qquad (8.3)$$

This angle is called the Prandtl Meyer angleand it is the angle through which a sonic flow has to be turned (away from itself) to reach a supersonic Mach number M. Alternatively, it is the angle through which a supersonic flow at a Mach number M has to be turned (towards itself) to reach $M = 1$. Both the expansion and the compression process are isentropic. Also note that v is a monotonically increasing function of M. Thus, if a supersonic flow is deflected through an angle θ in a compression corner, then $\theta = v_1 - v_2$. On the contrary, if it is deflected through the same angle in an expansion corner, then $\theta = v_2 - v_1$. Both the Prandtl-Meyer angle and the Mach angle are tabulated for Mach numbers ranging from 1 to 5 and $\gamma = 1.4$ in Table A.6.

Two limiting cases for v can be considered, namely, as $M \to 1$ and $M \to \infty$. In the former case, $v \to 0$ and $\mu \to 90°$, as already mentioned. In the latter case, it is easy to show from the above equation that,

$$v_{max} = \frac{\pi}{2}\left(\sqrt{\frac{\gamma + 1}{\gamma - 1}} - 1\right) \qquad (8.4)$$

and $\mu \to 0°$. For $\gamma = 1.4$, v_{max} is equal to $130.45°$. This maximum value is only of academic interest, since, long before the flow is deflected through this angle (and correspondingly expanded), the continuum assumption would have become invalid owing to the static pressure attaining very low values.

Values of μ and v are listed in Table A.6 for M rnaging from 1 to 5.

Example 8.1 Supersonic flow at $M = 3$, $P = 100$ kPa and $T = 300$ K is deflected through $20°$ at a compression corner. Determine the flow properties downstream of the corner, assuming the process to be isentropic.

Solution For $M_1 = 3$, the Prandtl Meyer angle $\nu_1 = 49.7568°$, from Table A.6. Since this is a compression corner,

$$\theta = \nu_1 - \nu_2 \text{ and so } \nu_2 = 29.7568°$$

From Table A.6, it can be seen that this value of ν corresponds to $M_2 = 2.125$. From the isentropic table,

$$T_2 = \frac{T_2}{T_0}\frac{T_0}{T_1} T_1 = \frac{0.5254575}{0.35714} \times 300 = 441 \text{ K}$$
$$P_2 = \frac{P_2}{P_0}\frac{P_0}{P_1} P_1 = \frac{36.7327}{9.509016} \times 100 = 386.3 \text{ kPa}$$

Comparison of these numbers with those given in Worked Example 7.1 reveals that the Mach number at the end of the compression process is higher and there is no loss of stagnation pressure now. Since the Mach number is higher, the static temperature is lower (since the stagnation temperature is the same in both the cases).

Example 8.2 Consider supersonic flow (M=3, P=100 kPa and T=300 K) around a 20° convex corner as shown in the figure. Determine the flow properties $(M, \theta, P, T,$ and $P_0)$ at $2s$ and 2. You may assume that 2 is sufficiently far away from the slip line.

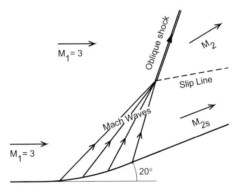

Solution Properties at state point $2s$ are the same as those calculated in the previous example. These are tabulated below.

Assuming that static pressure is continuous across the slip line (but other quantities need not be), we can proceed to determine the properties at state 2 as follows. Since $P_2/P_1 = 3.863$, we can retrieve from Table A.2, $M_{n,1} = 1.86$, $M_{n,2} = 0.6$ and $T_2/T_1 = 1.577$. Thus $T_2 = 473$ K.

From $M_{n,1} = M_1 \sin \beta_1$, we can get $\beta_1 = 38.32°$. From Table A.5, corresponding to these values for M_1 and β_1, we can get $\theta_1 = 20.46°$. With $M_{n,2} = M_2 \sin(\beta_1 - \theta_1)$, we can get $M_2 = 1.96$. For this value of M_2, we can get $P_{0,2}/P_2 = 7.35297$, from which $P_{0,2} = 2840.45$ kPa.

	M	θ	P	T	P_0
2s	2.125	20°	386.3 kPa	441 K	3673.27 kPa
2	1.96	20.46°	386.3 kPa	473 K	2840.45 kPa

8.4 Reflection of Oblique Shock from a Constant Pressure Boundary

A constant pressure boundary is a physically distinct surface in a fluid, across (and along) which the static pressure is the same. This type of boundary is usually encountered in the study of jets and mixing layers. As can be seen in Fig. 8.5, the surface of the jet is a constant pressure boundary. We now wish to find out what happens when the oblique shocks emanating from the top and bottom corners of the nozzle impinge on this boundary.

If we consider the point of impingement which is located right on the jet boundary, then there is an increase in pressure due to the shock wave. But the pressure at this point cannot be different from the ambient pressure and so the pressure rise due to the impingement of the shock wave is immediately nullified by the generation of an expansion fan. Thus, an oblique shock is reflected from a constant pressure boundary as an expansion fan. Consequently, the jet which was shrinking in diameter, begins to swell from this point onwards due to the expansion process.

By using the same argument as before for an oblique shock, it can be easily shown that expansion fans reflect from a wall as expansion fans, and they are reflected as weak oblique shock waves from a constant pressure boundary. In the latter case, the reflected weak shock waves eventually coalesce into an oblique shock of finite strength. This is illustrated in Fig. 8.5.

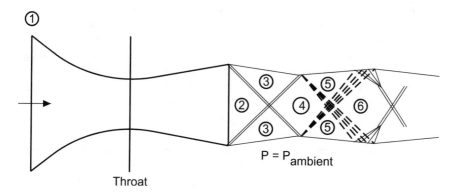

Fig. 8.5 Reflection of oblique shock and expansion fan from a constant pressure boundary

Example 8.3 An oblique shock impinges on a corner as shown in the figure. Determine the nature of the reflected wave (if any) in each case.

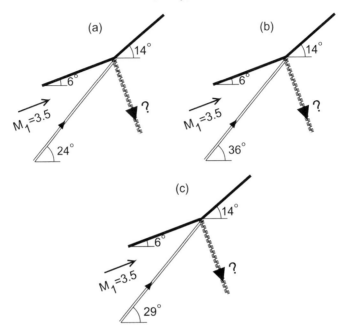

Solution (a) In this case, the wave angle $\beta_1 = 24° − 6° = 18°$. From Table A.5, for $M = 3.5$ and $\beta = 18°$, we can retrieve $\theta_1 = 2° = 8°$ counter-clockwise from the horizontal. This is the angle through which the impinging shock turns the flow. The angle through which the reflected wave has to turn the flow so that is parallel to the wall may be evaluated as $\theta_2 = 14° − 8° = 6°$ in the counter-clockwise direction. With respect to the direction of the reflected wave, since the flow has to be turned *away from* the wave, the reflected wave in this case is an **expansion fan**.

(b) In this case, the wave angle $\beta_1 = 36° − 6° = 30°$. From Table A.5, for $M = 3.5$ and $\beta = 30°$, we can retrieve $\theta_1 \approx 16° = 22°$ counter-clockwise from the horizontal. This is the angle through which the impinging shock turns the flow. The angle through which the reflected wave has to turn the flow so that is parallel to the wall may be evaluated as $\theta_2 = 22° − 14° = 8°$ in the clockwise direction. With respect to the direction of the reflected wave, since the flow has to be turned *towards* the wave, the reflected wave in this case is an **oblique shock**.

(c) In this case, the wave angle $\beta_1 = 29° − 6° = 23°$. From Table A.5, for $M = 3.5$ and $\beta = 23°$, we can retrieve $\theta_1 \approx 8° = 14°$ counter-clockwise from the horizontal. This is the angle through which the impinging shock turns the flow. The angle through which the reflected wave has to turn the flow so that is parallel to the wall may be evaluated as $\theta_2 = 14° − 14° = 0°$. In this case, the impinging shock wave is terminated at the corner and hence there is **no reflection**.

Example 8.4 Continue the worked example in Sect. 7.4 and determine the flow properties in region 5 (Fig. 8.5) and determine the angle made by the edge of the jet with the horizontal in this region.

Solution Across the expansion fan, the flow expands and reaches the ambient pressure, while the stagnation pressure remains constant. Thus, $P_5 = 100kPa$ and $P_{0,5} = 785.48$ **kPa**. From isentropic table, for $P_5/P_{0,5} = 0.127311$,

$$\mathbf{M_5} \approx \mathbf{2.0} \text{ and } \frac{T_5}{T_{0,5}} = 0.55556$$

Since $T_{0,5} = 300$ K, we get $\mathbf{T_5 = 167}$ **K**. From Table A.6, for $M_4 = 1.48$, $v_4 = 11.3168°$.

Similarly, $v_5 = 26.3795°$. Since this is an expansion process, flow deflection angle is $26.3795° - 11.3168° = 15.0627°$. The angle made by the edge of the jet with the horizontal is

$15.0627° + 14.6834° - 11.2118° = \mathbf{18.5343°}$ (counter clockwise)

Continuing the calculations beyond this point is somewhat difficult, as we need to look at the effect of the expansion fans intersecting each other. The reader may consult Zucrow and Hoffman for calculations involving such interactions.

Problems

8.1 Sketc.h the flow field for the flow through the intake shown in Fig. 7.5 indicating oblique shocks and expansion fans clearly. Also show the external flow field around the cowl. Assume critical mode of operation.

8.2 Air at a stagnation pressure of 1 MPa flows isentropically through a CD nozzle and exhausts into ambient at 40.4 kPa. The edge of the jet, as it comes out of the nozzle is deflected by 18° (counter clockwise) from the horizontal. Determine the Mach number and static pressure at the nozzle exit.
    [2.01, 125.8 kPa]

8.3 A sharp throated nozzle is shown in Fig. 8.6. The flow entering the throat is sonic. The exit to throat area ratio is 3 and the throat makes an angle of 45° with the horizontal. Determine the Mach number at the exit. Wave reflection may be ignored. The flow may be assumed to be quasi 1D, except near the corner. Also determine the exit Mach number for the same area ratio, if the throat were smooth.
    [3.944, 2.67]

8.4 A supersonic injector fabricated from a CD nozzle (comprising of a circular arc throat and a conical divergent portion) is shown in Fig. 8.7.[3] The throat diameter is

---

[3]This type of nozzle is encountered in steam and gas turbines also.

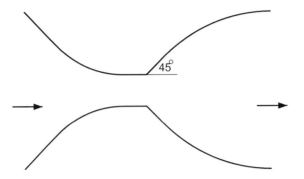

Fig. 8.6 A sharp throated nozzle

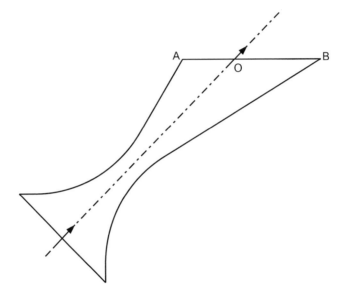

Fig. 8.7 An inclined supersonic injector

0.55 cm and the divergence angle is 10°. The axis of the nozzle is inclined at an angle
of 45° to the horizontal. The inlet stagnation temperature is 2400 K and the mass
flow rate is 0.005 kg/s. The static pressure at point O on the axis is the same as the
ambient pressure and is equal to 25 kPa. Determine (a) the Mach number at O, (b)
the static pressure and Mach number at points A and B and (c) the angle made by
the jet boundary with the axis at these points. Sketc.h the jet boundaries with respect
to the nozzle centerline for a few nozzle diameters as done in Fig. 8.5.

[2.17, 46 kPa, 1.775, 12.8 kPa, 2.6, 10.95° away from the axis, 10.52° towards
the axis]

Chapter 9
Flow of Steam Through Nozzles

In the earlier chapters, we studied the gas dynamics of a perfect gas. In this chapter, we will study the dynamics of the flow of steam through nozzles. Historically, the theory of the flow of steam through nozzles was developed first in the late 1800 and early 1900 s. The convergent divergent nozzle to accelerate steam to high speeds for use in impulse steam turbines was designed by de Laval in 1888. The importance of studying the flow of steam arises from the fact that steam turbines, even today, are used extensively in power generation. A clear understanding of the dynamics of the flow of steam through the blade passages of the steam turbines as well as through the nozzles which precede the blades is thus very important. The theory developed in Chap. 6 for calorically perfect gases can be carried over for steam with a few modifications, which will be mentioned shortly. First, a brief review of the thermodynamic states of water in the pressure and temperature range of interest is presented.

9.1 T-s Diagram of Liquid Water-Water Vapor Mixture

The thermodynamic state of a liquid water-water vapor mixture can be illustrated on a T-s diagram as shown in Fig. 9.1. The most salient feature in this diagram is the "dome" shaped region bounded by the curve ACB. The state of a mixture of liquid water and water vapor will lie inside this region. Curve CA is the saturated liquid line and curve CB is the saturated vapor line. The term saturated is used to highlight the fact that these lines indicate the beginning or termination of a change of phase. An examination of the isobars shown in this diagram makes it clear that the pressure and temperature remain constant when phase change takes place. For a given temperature T, the pressure at which phase change takes place is called the saturation pressure corresponding to that temperature, denoted as $P_{sat}(T)$. Alternatively, for a given pressure P, the temperature at which phase change takes place is called the saturation

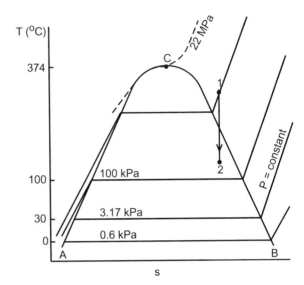

Fig. 9.1 T-s diagram for liquid water-water vapor mixture. Not to scale

temperature corresponding to that pressure and is denoted as $T_{sat}(P)$. States that lie to the left of CA and below C are referred to as compressed liquid states or sub-cooled states. This is because, for such states, the given (P, T) is such that $P > P_{sat}(T)$ or, alternatively, $T < T_{sat}(P)$. States that lie to the right of CB represent superheated vapor (steam). Point C (22 MPa and 374 °C) is called the critical state. To summarize, states that lie inside the dome region are two phase mixtures whereas states that lie outside correspond to a single phase (liquid or vapor).

Since pressure and temperature are not independent inside the two phase region, an additional property is required to fix the state. For this purpose, a new property called the dryness fraction (x) is introduced. This is defined as

$$x = \frac{m_g}{m} = \frac{m_g}{m_f + m_g} \tag{9.1}$$

where the subscripts f and g refer to the liquid and vapor phase respectively and m denotes the mass. It is easy to see that $x = 0$ corresponds to a saturated liquid and $x = 1$ corresponds to a saturated vapor state. Thus, the saturated liquid line CA is the locus of all states for which $x = 0$ and the saturated vapor line CB is the locus of states for which $x = 1$. The dryness fraction is indeterminate at the critical state C. Physically, this means that, when a substance exits at the critical state, it is impossible to distinguish whether it is in the liquid or vapor state. It is important to realize that the concept of dryness fraction is meaningless outside the dome region.

Any specific property of the two phase mixture can be evaluated as the weighted sum of the respective values at the saturated liquid and vapor states, where the weights can be expressed in terms x. For instance, let a two phase mixture of mass m contain m_f and m_g of saturated liquid and vapor respectively. The specific volume of the mixture (v) is given as

$$v = \frac{V}{m}$$

$$= \frac{m_f v_f + m_g v_g}{m}$$

$$= (1 - x)v_f + x v_g$$

$$= v_f + x(v_g - v_f)$$

Any specific property of the mixture (ϕ) in the two phase region can be written in the same manner as

$$\phi = \phi_f + x(\phi_g - \phi_f) \qquad (9.2)$$

where, ϕ can be the specific volume v, specific internal energy u, specific enthalpy h or the specific entropy s. Property values of superheated states can be retrieved either from tables or the Mollier diagram ($h - s$ diagram). We mention in passing that, in contrast to the earlier chapters, here, we will use u to denote the internal energy instead of e.

9.2 Isentropic Expansion of Steam

In the earlier chapters, the development assumed the fluid to be a calorically perfect gas. It may be recalled that this model assumes that

- the gas obeys the ideal gas equation of state, $Pv = RT$ and
- the internal energy is a linear function of temperature.

In the case of steam (as well as a two phase saturated mixture of liquid and vapor), both these assumptions are unrealistic and must be abandoned. The T-v diagram shown in Fig. 9.2 illustrates how well superheated steam obeys the ideal gas equation of state. Thermodynamic states that lie within the shaded region in this figure are such that $Pv/(RT) - 1 \leq 1$ percent. States that lie outside this region show greater departure from ideal gas behaviour. States near the critical point, C, show the maximum departure. In an actual application, there is no guarantee that the initial state of the steam (before expansion in the nozzle) will lie within the shaded region in Fig. 9.2. More importantly, the state at the end of the isentropic expansion process in the nozzle (process 1–2 in Figs. 9.1 and 9.2) will most likely lie in the two phase region.

The internal energy of water in the superheated region is a function of temperature and pressure, i.e., $u = u(T, P)$ and a function of temperature and specific volume in the two-phase region, i.e., $u = u(T, v)$. Hence, even the thermally perfect assumption is invalid let alone the calorically perfect assumption.

In view of the arguments presented above, actual calculations involving expansion of steam in nozzles and blade passages have to be carried out using tabulated property

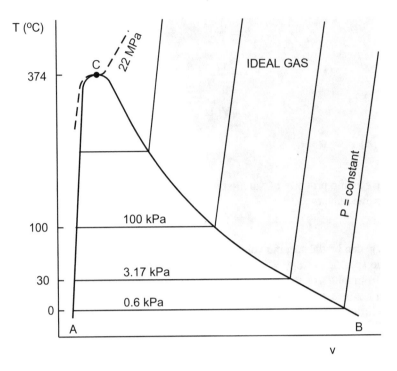

Fig. 9.2 T-v diagram for liquid water-water vapor mixture. Not to scale. *Adapted from Thermodynamics by Cengel and Boles, McGraw Hill, Fifth edition*

data or the Mollier diagram. However, it would be convenient, to the extent possible, to be able to use closed form expressions as was done for ideal gases. This is explored next.

When steam undergoes an isentropic expansion process in a nozzle, if the process line crosses the saturated vapor line as shown in Figs. 9.1 and 9.2, condensation takes place and the steam becomes a two phase mixture of liquid and vapor. The nucleation and formation of the liquid droplets is a complex process and depends, among other things, upon the purity of the steam. Depending upon their radii, the droplets will tend to move with velocities different from the vapor phase owing to inertia force. Furthermore, as mentioned in Sect. 6.7, normal shocks tend to occur in the divergent portion of a convergent divergent nozzle. At this point, the steam would have undergone a considerable amount of expansion and is likely to be quite wet. As a result of the increase in pressure, temperature and entropy across the normal shock, the steam will become dry and saturated or even superheated. The gas dynamics of a two phase mixture is complex and beyond the scope of this book. However, if the dryness fraction does not become too low i.e., the steam does not become too wet, then the aforementioned complications can be ignored. We assume this to be the case in the rest of this chapter. In addition, we will assume that the flow is shock free and hence isentropic.

We know that isentropic expansion of an ideal gas obeys $Pv^\gamma = \text{constant}$. Isentropic expansion of steam can be represented using a similar expression of the form

$$Pv^n = \text{constant} \tag{9.3}$$

where the exponent n has to be determined from experimental data through a curve fit. This is given as

$$n = \begin{cases} 1.3 & \text{if superheated} \\ 1.035 + 0.1\,x_1 & \text{saturated mixture} \end{cases} \tag{9.4}$$

where x_1 is the *initial* dryness fraction. The exponent for superheated steam is due to Callendar and the expression for the saturated mixture is due to Zeuner. It will be demonstrated later through numerical examples that Eq. 9.3 with the exponent given by Eq. 9.4 is an excellent description of the isentropic expansion process of steam in nozzles.

Since the propagation of an acoustic wave through a medium causes changes in the properties of the medium that are governed by an isentropic process, the speed of sound a, in superheated steam or a saturated mixture can be calculated using Eqs. 2.13 and 9.4 as follows:

$$a = \sqrt{\left.\frac{dP}{d\rho}\right|_s} f \tag{2.13}$$

$$= \sqrt{nPv} \tag{9.5}$$

9.3 Flow of Steam Through Nozzles

The theory developed in Chap. 6 for the flow of a calorically perfect gas is applicable for the flow of steam as well, subject to the constraints mentioned above. In view of this, this material will not be repeated in this chapter. However, the material will be developed along the lines customarily adopted in the context of steam nozzles.

The continuity equation for this flow can be written in differential form as

$$d(\rho V A) = 0 \tag{6.3}$$

where, in contrast to the earlier chapters, we have used V to denote the velocity.[1] Momentum and energy equation in differential form are the same as Eq. 2.2 and 2.3, *viz.,*

$$dP + \rho V dV = 0 \tag{2.2}$$

---

[1] Since we have used u to denote the internal energy now.

$$dh + d\left(\frac{V^2}{2}\right) = 0 \tag{2.3}$$

If we integrate the momentum equation, we get, for expansion between states 1 and 2,

$$\frac{V_2^2 - V_1^2}{2} = -\int_1^2 v \, dP$$

Since $Pv^n = \text{constant}$, the integral on the right hand can be evaluated. This leads to

$$\frac{V_2^2 - V_1^2}{2} = \frac{n}{n-1}\left(P_1 v_1 - P_2 v_2\right)$$

This can be rewritten as

$$V_2 = \sqrt{V_1^2 + \frac{2n}{n-1} P_1 v_1 \left[1 - \frac{P_2 v_2}{P_1 v_1}\right]}$$

$$= \sqrt{V_1^2 + \frac{2n}{n-1} P_1 v_1 \left[1 - \left(\frac{P_2}{P_1}\right)^{(n-1)/n}\right]} \tag{9.6}$$

where we have used $P_1 v_1^n = P_2 v_2^n$. If the steam expands from a steam chest, where stagnation conditions prevail, then $V_1 = 0$, $P_1 = P_0$ and $v_1 = v_0$. Thus,

$$V_2 = \sqrt{\frac{2n}{n-1} P_0 v_0 \left[1 - \left(\frac{P_2}{P_0}\right)^{(n-1)/n}\right]}$$

If we integrate Eq. 2.3 between states 1 and 2, we get

$$h_1 - h_2 = \frac{V_2^2 - V_1^2}{2}$$

Upon substituting for V_2 from Eq. 9.6, we get

$$h_1 - h_2 = \frac{n}{n-1} P_1 v_1 \left[1 - \left(\frac{P_2}{P_1}\right)^{(n-1)/n}\right] \tag{9.7}$$

Without any loss of generality, if we simply assume that the steam is expanded from the chest to a final pressure of P, then the subscript 2 can be dropped from the above equation and we thus get the velocity at the end of the expansion process to be

$$V = \sqrt{\frac{2n}{n-1} P_0 v_0 \left[1 - \left(\frac{P}{P_0}\right)^{(n-1)/n}\right]} \tag{9.8}$$

In contrast to nozzle flows involving perfect gases where the Area-Mach number relation is used extensively (and perhaps exclusively), Eq. 9.8 is used extensively in applications involving flow of steam through nozzles.[2]

9.3.1 Choking in Steam Nozzles

The mass flow rate at any section in a nozzle is given as

$$\dot{m} = \frac{AV}{v}$$

If we use the fact that $Pv^n = P_0 v_0^n$ in the above expression, we get

$$\dot{m} = \frac{AV}{v_0} \left(\frac{P}{P_0}\right)^{1/n}$$

Upon substituting for V from Eq. 9.8, we are led to

$$\dot{m} = A \sqrt{\frac{2n}{n-1} \frac{P_0}{v_0} \left[\left(\frac{P}{P_0}\right)^{2/n} - \left(\frac{P}{P_0}\right)^{(n+1)/n}\right]}$$

For a given value of A, P_0 and v_0, it is easy to show that the mass flow rate is a maximum when the steam is expanded to a pressure P given by

$$\frac{P}{P_0} = \left(\frac{2}{n+1}\right)^{n/(n-1)} \tag{9.9}$$

We can rewrite Eq. 9.8 as

$$V = \sqrt{\frac{2n}{n-1} Pv \left[\left(\frac{P_0}{P}\right)^{n/(n-1)} - 1\right]}$$

If we substitute for P from Eq. 9.9, then, at the section where the pressure is given by this equation, the velocity of the steam is given as

$$V = \sqrt{nPv}$$

It follows from Eq. 9.5 that this velocity is equal to the local speed of sound. Hence, Eq. 9.9 may be rewritten as

---

[2]Most likely due to the almost exclusive use of steam nozzles in steam turbines.

$$\frac{P^*}{P_0} = \left(\frac{2}{n+1}\right)^{n/(n-1)} \tag{9.10}$$

It is worthwhile emphasizing that Eqs. 9.8, 9.9 and 9.10 have been derived with the sole assumption that the isentropic expansion process of steam in the nozzle can be described by $Pv^n = \text{constant}$ and without using the calorically perfect assumption.

Example 9.1 Dry, saturated steam enters a convergent nozzle at a static pressure of 500 kPa and is expanded to 300 kPa. If the inlet and throat diameters are 0.05 m and 0.025 m respectively, determine the velocity at the inlet and exit and the stagnation pressure.

Solution Given $P_1 = 500$ kPa and $x_1 = 1$, we can get $v_1 = 0.3748\,\text{m}^3/\text{kg}$, $h_1 = 2748.49$ kJ/kg and $s_1 = 6.8215$ kJ/kg.K from Table A.8. Since the steam is initially dry and saturated, $n = 1.135$ from Eq. 9.4. From Eq. 9.3, we can get

$$v_e = v_1 \left(\frac{P_e}{P_1}\right)^{-1/n} = 0.5878\,\text{m}^3/\text{kg}$$

and from Eq. 9.7

$$h_1 - h_e = \frac{n}{n-1} P_1 v_1 \left[1 - \left(\frac{P_e}{P_1}\right)^{(n-1)/n}\right] = 92.88\,\text{kJ/kg}$$

$$= \frac{V_e^2 - V_1^2}{2}$$

Also, since the mass flow rates at the inlet and outlet are the same, we have

$$V_1 = \frac{A_e}{A_1} \frac{v_1}{v_e} V_e$$

(a) We can thus obtain the exit velocity $V_e = 436.6\,\text{m/s}$. It follows that the inlet velocity $V_1 = 69.6\,\text{m/s}$. Note that the speed of sound at the exit can be calculated from Eq. 9.5 to be 447.4 m/s. Hence, the flow is not choked.

(b) The stagnation enthalpy can be evaluated using

$$h_0 = h_1 + \frac{V_1^2}{2} = 2751\,\text{kJ/kg}$$

Since $s_0 = s_1 = 6.8215$ kJ/kg.K, we can get the stagnation pressure $P_0 = 507$ kPa from the steam table. For this value of P_0, we can obtain $P^* = 294$ kPa from Eq. 9.10. The given exit pressure is higher than this value confirming our earlier observation that the flow is not choked.

Example 9.2 Dry, saturated steam at 1 MPa in a steam chest expands through a nozzle to a final pressure of 100 kPa. Determine (a) if the nozzle is convergent or convergent-divergent, (b) the exit velocity, (c) the dryness fraction at the exit and (c) the exit to throat area ratio. Assume the expansion to be isentropic throughout.

Solution We have $P_0 = 1$ MPa and we can get $v_0 = 0.19436 \, \text{m}^3/\text{kg}$ from Table A.8.

(a) Since the steam is initially dry and saturated, $n = 1.135$ from Eq. 9.4. From Eq. 9.10, we have $P^* = 0.58 \, P_0 = 580$ kPa. Since the flow is expanded to a final pressure of 100 kPa in the nozzle, it is clear that the nozzle is convergent-divergent.

(b) We have from Eq. 9.7

$$ h_0 - h_e = \frac{n}{n-1} P_0 v_0 \left[1 - \left(\frac{P_e}{P_0} \right)^{(n-1)/n} \right] $$

where the subscript e denotes the exit. Upon substituting numerical values, we get

$$ h_0 - h_e = 391.48 \, \text{kJ/kg} $$

Therefore

$$ V_e = \sqrt{2(h_0 - h_e)} = 885 \, \text{m/s} $$

(c) From Eq. 9.3, we can get

$$ v_e = v_0 \left(\frac{P_e}{P_0} \right)^{-1/n} = 1.478 \, \text{m}^3/\text{kg} $$

Using steam tables, the dryness fraction at the exit can now be calculated as 0.87.

(d) From Eq. 9.3, we can get

$$ v^* = v_0 \left(\frac{P^*}{P_0} \right)^{-1/n} = 0.3141 \, \text{m}^3/\text{kg} $$

and

$$ V^* = \sqrt{n P^* v^*} = 454.7 \, \text{m/s} $$

Since

$$ \dot{m} = \frac{A^* V^*}{v^*} = \frac{A_e V_e}{v_e} $$

we can obtain $A_e/A^* = 2.42$.

An alternative solution method is to use the Mollier diagram or the steam table. From the steam table, for dry saturated vapor at 1 MPa, we can get $h_0 = 2778.1$ kJ/kg and $s_0 = 6.5865$ kJ/kg.K. Since the expansion process is isentropic, we have at the exit, $s_e = 6.5865$ kJ/kg.K and $P_e = 100$ kPa. From steam tables, we can get $x_e = 0.872$ and thus $h_e = 2387.3$ kJ/kg and $v_e = 1.4773$ m³/kg. Therefore

$$V_e = \sqrt{2(h_0 - h_e)} = 884 \text{ m/s}$$

Example 9.3 Steam at 700 kPa, 250°C in a steam chest expands through a nozzle to a final pressure of 100 kPa. The mass flow rate is 0.076 kg/s. Determine (a) if the nozzle is convergent or convergent-divergent (b) the throat diameter (c) the exit diameter and (d) the dryness fraction at the exit. Assume the expansion process to be isentropic and in equilibrium throughout.

Solution From Table A.9, it is easy to establish that for the given steam chest conditions of $P_0 = 700$ kPa and $T_0 = 250$°C the steam is initially superheated. Also, $v_0 = 0.336343$ m³/kg and $s_0 = 7.1062$ kJ/kg.K.

(a) Since the steam is initially superheated, $n = 1.3$ from Eq. 9.4. From Eq. 9.10, we have $P^* = 0.545 P_0 = 380$ kPa. Since the flow is expanded to a final pressure of 100 kPa in the nozzle, it is clear that the nozzle is convergent-divergent.

(b) From Eq. 9.3, we can get

$$v^* = v_0 \left(\frac{P^*}{P_0}\right)^{-1/n} = 0.53648 \text{ m}^3/\text{kg}$$

and

$$V^* = \sqrt{n P^* v^*} = 514.8 \text{ m/s}$$

Since

$$A^* = \frac{\dot{m} v^*}{V^*} = 7.92 \times 10^{-5} \text{ m}^2$$

the throat diameter $D^* = 10$ mm.

(c) Since the expansion process is isentropic, the process line crosses the saturated vapor line at $P_g = 212$ kPa. With $n = 1.3$ (as the flow is superheated until it crosses the saturated vapor line), we can get

$$h_0 - h_g = \frac{n}{n-1} P_0 v_0 \left[1 - \left(\frac{P_g}{P_0}\right)^{(n-1)/n}\right] = 245.8 \text{ kJ/kg}$$

and

$$v_g = v_0 \left(\frac{P_g}{P_0}\right)^{-1/n} = 0.843 \text{ m}^3/\text{kg}$$

It follows that $V_g = \sqrt{2(h_0 - h_g)} = 701.14$ m/s. Once the flow crosses the saturated vapor line, $n = 1.135$ (this implies that the flow continues to be in equilibrium after crossing the saturated vapor line). Hence

$$h_g - h_e = \frac{n}{n-1} P_g v_g \left[1 - \left(\frac{P_e}{P_g} \right)^{(n-1)/n} \right] = 127.77 \,\text{kJ/kg}$$

and

$$v_e = v_g \left(\frac{P_e}{P_g} \right)^{-1/n} = 1.6344 \,\text{m}^3/\text{kg}$$

Therefore

$$V_e = \sqrt{V_g^2 + 2(h_g - h_e)} = 864 \,m/s$$

For the given mass flow rate, the exit area can be calculated as

$$A_e = \frac{\dot{m} v_e}{V_e} = 1.44 \times 10^{-4} \,\text{m}^2$$

and the exit diameter can be calculated as $D_e = 13.5$ mm.

(d) From the given value of the exit pressure P_e and the calculated value of v_e, we can get the dryness fraction at the exit to be 0.96 using steam tables.

Alternatively, we can use the steam table (or Mollier diagram) to solve the problem. For the given steam chest condition, we can get $h_0 = 2954$ kJ/kg and $s_0 = 7.1062$ kJ/kg.K.

The pressure at the throat $P^* = 0.545\, P_0 = 380$ kPa. Since the expansion is isentropic, $s^* = s_0 = 7.1062$ kJ/kg.K. Therefore, the fluid is superheated at the throat with $v^* = 0.537$ m³/kg and $h^* = 2820$ kJ/kg. It follows that

$$V^* = \sqrt{2 (h_0 - h^*)} = 518 \,m/s$$

and

$$A^* = \frac{\dot{m} v^*}{V^*} = 7.86 \times 10^{-5} \,\text{m}^2$$

Thus, the throat diameter $D^* = 10$ mm.

We have $P_e = 100$ kPa and $s_e = s_0 = 7.112$ kJ/kg.K. It can be determined that the fluid at the exit is a saturated mixture. The dryness fraction x_e can be evaluated as 0.96. Also, $h_e = 2581$ kJ/kg and $v_e = 1.623$ m³/kg. Hence

$$V_e = \sqrt{2 (h_0 - h_e)} = 864 \,m/s$$

and

$$A_e = \frac{\dot{m} v_e}{V_e} = 1.43 \times 10^{-4} \,\text{m}^2$$

Thus, the exit diameter $D_e = 13.5$ mm.

9.4   Supersaturation and the Condensation Shock

When thermodynamic states and processes are depicted in diagrams such as the ones in Figs. 9.1 and 9.2, an important assumption is that equilibrium prevails during the process. In other words, the underlying implication is that the system is given sufficient time to attain that state i.e., the process takes place so slowly that the system moves from one equilibrium state to another. While this can be taken for granted in most situations, it needs to be re-examined when the process takes place very rapidly such as during the expansion in nozzles.

The numerical examples above illustrate that the steam attains velocities of the order of several hundred metres per second as it flows through a nozzle (convergent or convergent-divergent). Consequently, the expansion process may be out of equilibrium at high velocities, since the fluid is not allowed enough time to attain the successive thermodynamic states. This means that the actual state of the fluid at a certain pressure and specific volume at a location in the nozzle will not coincide with the state on the T-s diagram corresponding to the same pressure and specific volume. This departure from equilibrium is more pronounced when the process line crosses the saturated vapor curve from the superheated region into the two phase region. This is owing to the fact that condensation requires a finite amount of time for droplets to form and grow.

It is clear from the isentropic expansion process shown in Fig. 9.3 that the vapor, starting from a superheated state would have expanded considerably and would be moving at a high velocity when the process line crosses the saturation curve.[3] Hence the flow is likely to be out of equilibrium (or, alternatively, in a metastable equilibrium) at or beyond the throat. Consequently, condensation is delayed and the vapor continues to exist and expand as a vapor. For instance, if the fluid were to be at equilibrium at state x' in Fig. 9.3, it would exist as a two phase mixture at a pressure $P_{x'}$ and the corresponding equilibrium temperature $T_{sat}(P_{x'})$. However, if it were to be out of equilibrium at the same pressure, then it would exist at state x as a vapor at pressure $P_x = P_{x'}$ and temperature T_x. Since the expansion process is isentropic, state point x' is located at the point of intersection of the isentrope $s = s_0$ and the isobar $P = P_{x'}$. Whereas, state point x is located at the intersection of the same isentrope and the portion of the isobar $P = P_{x'}$ in the superheated region extended into the two phase region. Thus, the temperature at state x is not $T_{sat}(P_x)$ but is given as

$$T_x = T_0 \left(\frac{P_x}{P_0} \right)^{(n-1)/n} \tag{9.11}$$

where we have used the fact that $P_x = P_{x'}$. It must be noted that in writing Eq. 9.11 from Eq. 9.3, we have used the ideal gas equation of state for the supersaturated vapor. This is only an approximation as evident from Fig. 9.2, but an acceptable one for practical purposes.

---

[3]Even if the steam starts out as dry, saturated vapor, it would still acquire a high velocity within a short distance inside the nozzle.

Fig. 9.3 Illustration of supersaturated flow on a T-s diagram. Note that x is a metastable state and the dashed line connecting $x - y$ represents the condensation shock

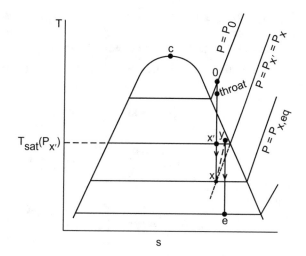

Two quantities are customarily used to characterize supersaturated steam. These are the degree of supersaturation or supersaturation ratio denoted by S and the degree of supercooling. The degree of supersaturation is defined as the ratio of the pressure P of the supersaturated vapor to the (equilibrium) saturation pressure corresponding to its temperature. Thus,

$$S = \frac{P}{P_{sat}(T)} \tag{9.12}$$

The degree of supercooling of a supersaturated vapor is defined as the difference between the actual temperature T and the (equilibrium) saturation temperature corresponding to its pressure. Thus,

$$\Delta T = T_{sat}(P) - T \tag{9.13}$$

Figure 9.3 illustrates how these quantities are to be evaluated. It should also be clear from this figure that $S > 1$ and $\Delta T > 0$ for a supersaturated vapor. The degree of supercooling is the exact opposite of the more familiar degree of superheat. It may be recalled that the latter quantity is defined in the same manner except that $T > T_{sat}(P)$.

When steam exists in this unnaturally dry, supersaturated state, its density is higher than that of the saturated vapor at the same pressure by a factor of approximately 5–8. Moreover, since the latent heat of condensation has not been released owing to the delayed condensation, the enthalpy drop and hence the velocity (Eq. 9.7) is also less. This reduction is usually not high as the velocity is dependent on the square root of the enthalpy drop. However, if the flow were to be supersaturated at the nozzle throat, then the combined effect of these two factors is to **increase** the mass flow rate through the nozzle for a given steam chest pressure and throat area.

Example 9.4 For the same steam chest condition and exit pressure in the previous example, determine the exit velocity, the supersaturation ratio and the degree of supercooling, if the flow is out of equilibrium.

Solution In this case, we use $n = 1.3$ for the entire expansion process. Therefore,

$$v_e = v_0 \left(\frac{P_e}{P_0} \right)^{-1/n} = 4.2209 \, \text{m}^3/\text{kg}$$

$$T_e = T_0 \left(\frac{P_e}{P_0} \right)^{(n-1)/n} = 333.8 \, K = 60.8 \, ^\circ C$$

and

$$h_0 - h_e = \frac{n}{n-1} P_0 v_0 \left[1 - \left(\frac{P_e}{P_0} \right)^{(n-1)/n} \right] = 369.1 \, \text{kJ/kg}$$

The exit velocity $V_e = \sqrt{2(h_0 - h_e)} = 859$ m/s.

The saturation pressure corresponding to $T_e = 60.8\,^\circ C$ is 20.7 kPa. Hence, the supersaturation ratio $S = P_e/P_{sat}(T_e) = 4.83$.

The saturation temperature corresponding to $P_e = 100$ kPa is $100\,^\circ C$. Hence, the degree of supercooling $\Delta T = T_{sat}(P_e) - T_e = 39.2\,^\circ$ C.

As the supersaturated steam expands, both the degree of supersaturation and the degree of supercooling increase. However, there is a limit to the degree of supersaturation that can be allowed and once this limit is reached, the supersaturated vapor condenses almost instantaneously, triggering a *condensation shock* (Fig. 9.4). The pressure, temperature and entropy increase across the condensation shock as shown in Fig. 9.3. The limiting value of the degree of supersaturation is 5 and it depends upon, among other things, the purity of the vapor. If impurities such as dissolved salts are present, then nucleation of droplets begins early. Remarkably, experimental evidence suggests that the condensation shock is almost always initiated once the process line reaches the 96% dryness fraction line. This line is called the Wilson line. The term "shock" is somewhat of a misnomer since the condensation process occurs over a small but finite distance and is not a discontinuity.

The flow attains equilibrium downstream of the condensation shock and may continue to expand until the nozzle exit (denoted as the state point e in Fig. 9.3). However, if the back pressure at the exit is high, then a normal shock will stand somewhere in the divergent part of the nozzle as mentioned in Sect. 6.7. As before, the pressure, temperature and entropy increase across the shock wave. As a result, the two phase mixture upstream of the shock wave may attain the saturated vapor state or even become superheated downstream. This is shown in Fig. 9.4.

Condensation shocks are also seen in supersonic wind tunnels and nozzles that utilize air as the working fluid owing to the small amount of moisture that is present in the air. This moisture exists as a superheated vapor at its partial pressure and same temperature as the air. As the air expands and accelerates, so does the water vapor.

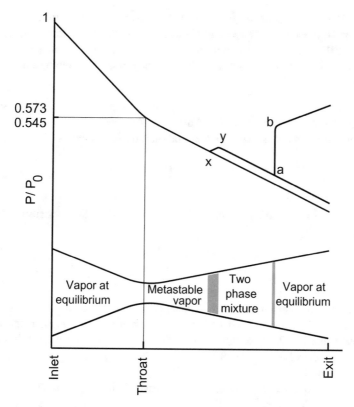

Fig. 9.4 Variation of static pressure along the nozzle. State points *x, y* lie across a condensation shock, while state points *a, b* lie across an aerodynamic (normal) shock

Similar to what happens in a steam nozzle, the water vapor becomes supersaturated and a condensation shock forms once the supersaturation limit is reached. The only difference in this case is that the water vapor is carried along by the air.

The development in Sects. 9.2 and 9.3 cannot be used downstream of state point *x* shown in Fig. 9.4, as two phase effects are significant. A comprehensive and unified theory of condensation as well as aerodynamic (normal) shocks in flows with or without a carrier gas was developed by Guha.[4] Interested readers may refer this work for the details of how to incorporate two phase effect and droplet nucleation and growth effect into the quasi one dimensional theory developed above in Sects. 9.2 and 9.3.

---

[4] "A unified theory of aerodynamic and condensation shock waves in vapor-droplet flows with or without a carrier gas", Physics of Fluids, Vol. 6, No. 5, May 1994, pp. 1893-1913.

Example 9.5 Steam expands in a convergent-divergent nozzle from a stagnation pressure and temperature of $P_0 = 144$ kPa and $T_0 = 118.7\,°C$ respectively. The steam expands until a supersaturation ratio of 5 is reached at which point a condensation shock occurs. Determine the pressure, velocity and degree of supercooling just ahead of the condensation shock.

Solution It can be established that, for the given stagnation conditions, the steam is superheated at the inlet and $v_0 = 1.234$ m^3/kg. Let x denote the state point just ahead of the condensation shock. The supersaturation ratio is defined as

$$S = \frac{P_x}{P_{sat}(T_x)}$$

If the flow is assumed to be isentropic until the onset of the condensation shock, then

$$T_x = T_0 \left(\frac{P_x}{P_0}\right)^{(n-1)/n} \tag{9.14}$$

with $n = 1.3$ for superheated steam.

The task is to determine P_x and T_x from these two equations for the prescribed value of S. This can be accomplished quite easily in an iterative manner starting with a guessed value for P_x. The initial guess should be made keeping in mind that P_x should be less than $P_{throat} = 0.58\,P_0 = 78584$ Pa. The procedure is illustrated in the following Table 9.1.

The calculations are carried out with progressively better guesses for P_x until the change in the value of P_x is less than 1 Pa. For this case, it can be seen from above that P_x converges to a value of 62691 Pa.

Table 9.1 Calculation of P_x

P_x (Pa)	$T_x°$ C	$P_{sat}(T_x)$ (Pa)	S
70000	58.63608141	18720.06507	3.739303242
60000	47.04608261	10650.71912	5.633422431
65000	53.0127163	14320.40686	4.538977184
62500	50.07530998	12397.50351	5.041337553
62700	50.31359558	12544.76681	4.998100081
62650	50.25407906	12507.84511	5.008856398
62670	50.27789005	12522.60531	5.004549648
62680	50.28979336	12529.98965	5.002398386
62690	50.3016952	12537.37682	5.00024853
62695	50.30764557	12541.07146	4.99917413
62691	50.3028853	12538.11569	**5.000033622**
62692	50.30407539	12538.85459	4.999818728

The enthalpy drop $h_0 - h_x$ can be evaluated from

$$h_0 - h_x = \frac{n}{n-1} P_0 v_0 \left[1 - \left(\frac{P_x}{P_0} \right)^{(n-1)/n} \right] = 134.46 \, \text{kJ/kg}$$

The velocity just ahead of the condensation shock can now be calculated as

$$V_x = \sqrt{2(h_0 - h_x)} = 518.6 \, \text{m/s}$$

From steam tables, we can get $T_{sat}(P_x) = 87\,^\circ\text{C}$. Hence, the degree of supercooling $\Delta T = T_{sat}(P_x) - T_x = 36.7\,^\circ\text{C}$.[5]

Problems

9.1 Dry, saturated steam enters a convergent nozzle at a static pressure of 800 kPa and is expanded to the sonic state. If the inlet and throat diameters are 0.05 m and 0.025 m respectively, determine the velocity at the inlet and exit and the stagnation pressure.

[70.6 m/s, 453 m/s, 806 kPa]

9.2 Dry saturated steam at 1.1 MPa is expanded in a nozzle to a pressure of 15 kPa. Assuming the expansion process to be isentropic and in equilibrium throughout, determine (a) if the nozzle is convergent or convergent-divergent, (b) the exit velocity, (c) the dryness fraction at the exit and (c) the exit to throat area ratio.

[convergent-divergent, 1146 m/s, 0.8, 11]

9.3 Superheated steam at 700 kPa, 220 °C in a steam chest is expanded through a nozzle to a final pressure of 20 kPa. The throat diameter is 10 mm. Assuming the expansion process to be isentropic and in equilibrium throughout, determine (a) the mass flow rate, (b) the exit velocity, (c) the dryness fraction at the exit and (c) the exit diameter.

[0.0781 kg/s, 1085 m/s, 0.87, 24.5 mm]

9.4 Dry saturated steam at 1.2 MPa is expanded in a nozzle to 20 kPa. The throat diameter of the nozzle is 6 mm. If the total mass flow rate is 0.5 kg/s, determine how many nozzles are required and the exit diameter of the nozzle. Assume the expansion process to be isentropic and in equilibrium throughout.

[10,0.18.2 mm]

9.5 Superheated steam at 850 kPa, 200 °C expands in a convergent nozzle until it becomes a saturated vapor. Determine the exit velocity, assuming the expansion process to be isentropic and in equilibrium throughout.

[407 m/s]

---

[5] See "A Simple and Fast Method for Calculating Properties Across a Condensation Shock", *ASME Journal of Fluids Engineering*, Vol. 140, Oct. 2018.

9.6 Steam which is initially saturated and dry expands from 1400 kPa to 700 kPa. Assuming the expansion to be in equilibrium (n = 1.135), determine the final velocity and specific volume. If the expansion is out of equilibrium (n = 1.3), determine the final velocity, specific volume, supersaturation ratio and the degree of undercooling.

[512 m/s, 0.2593 m³/kg; 502 m/s, 0.24 m³/kg, 3.41, 39°C]

9.7 Superheated steam at 500 kPa, 180°C is expanded in a nozzle to pressure of 170 kPa. Assuming the expansion process to be isentropic and in equilibrium determine the exit velocity. Assuming the flow to be isentropic and supersaturated, determine the the exit velocity, supersaturation ratio and the degree of supercooling.

[627 m/s, 621 m/s, 3.59, 34°C]

9.8 For each of the stagnation condition given below, determine the pressure, velocity and degree of supercooling just before the onset of condensation shock for a limiting value of supersaturation ratio of 5. Assume the expansion process to be isentropic. (a) 87000 Pa, 96°C, (b) 70727 Pa, 104°C and (c)25000 Pa, 85°C.

[45037 Pa, 453 m/s, 35°C; 29918 Pa, 518.6 m/s, 33°C; 10127 Pa, 518 m/s, 28°C]

Suggested Readings

Gas Dynamics Vol. 1 by M. J. Zucrow and J. D. Hoffman, John Wiley & Sons, 1976.
Dynamics and Thermodynamics of Compressible Fluid Flow by A. H. Shapiro, Krieger Publications Company, Reprint Edition, 1983.
Elements of Gas Dynamics by H. W. Liepmann and A. Roshko, Dover Publications, 2002.
Modern Compressible Flow With Historical Perspective by J. D. Anderson, Third Edition, McGraw-Hill Series in Aeronautical and Aerospace Engineering, 2003.
Compressible Fluid Dynamics by B. K. Hodge and K. Koenig, Prentice Hall Inc, 1995.
Elements of Gas Turbine Propulsion by J. D. Mattingly, McGraw-Hill Series in Aeronautical and Aerospace Engineering, 1996.
Compressible Fluid Flow by M. Saad, Second Edition, Prentice Hall Inc, 1985.
Fluid Mechanics by F. M. White, Fifth Edition, McGraw-Hill Higher Education, 2003.
Steam Turbine Theory and Practice by W. J. Kearton, Seventh Edition, Sir Isaac Pitman & Sons Ltd, London, 1958. Reprinted with permission in India by CBS Publishers and Distributors Pvt. Ltd, 2004.
Principles of Turbomachinery by S. A. Korpela, First Edition, John Wiley & Sons, 2011.

Tables A.1, A.2, A.3, A.4, A.5, A.6, A.7, A.8, and A.9.

© The Author(s) 2021
V. Babu, *Fundamentals of Gas Dynamics*,
https://doi.org/10.1007/978-3-030-60819-4

Table A.1 Isentropic table for $\gamma = 1.4$

M	$\frac{T_0}{T}$	$\frac{P_0}{P}$	$\frac{\rho_0}{\rho}$	$\frac{A}{A^*}$
0.00	1.00000E+00	1.00000E+00	1.00000E+00	∞
0.01	1.00002E+00	1.00007E+00	1.00005E+00	57.87384
0.02	1.00008E+00	1.00028E+00	1.00020E+00	28.94213
0.03	1.00018E+00	1.00063E+00	1.00045E+00	19.30054
0.04	1.00032E+00	1.00112E+00	1.00080E+00	14.48149
0.05	1.00050E+00	1.00175E+00	1.00125E+00	11.59144
0.06	1.00072E+00	1.00252E+00	1.00180E+00	9.66591
0.07	1.00098E+00	1.00343E+00	1.00245E+00	8.29153
0.08	1.00128E+00	1.00449E+00	1.00320E+00	7.26161
0.09	1.00162E+00	1.00568E+00	1.00405E+00	6.46134
0.10	1.00200E+00	1.00702E+00	1.00501E+00	5.82183
0.11	1.00242E+00	1.00850E+00	1.00606E+00	5.29923
0.12	1.00288E+00	1.01012E+00	1.00722E+00	4.86432
0.13	1.00338E+00	1.01188E+00	1.00847E+00	4.49686
0.14	1.00392E+00	1.01379E+00	1.00983E+00	4.18240
0.15	1.00450E+00	1.01584E+00	1.01129E+00	3.91034
0.16	1.00512E+00	1.01803E+00	1.01285E+00	3.67274
0.17	1.00578E+00	1.02038E+00	1.01451E+00	3.46351
0.18	1.00648E+00	1.02286E+00	1.01628E+00	3.27793
0.19	1.00722E+00	1.02550E+00	1.01815E+00	3.11226
0.20	1.00800E+00	1.02828E+00	1.02012E+00	2.96352
0.21	1.00882E+00	1.03121E+00	1.02220E+00	2.82929
0.22	1.00968E+00	1.03429E+00	1.02438E+00	2.70760
0.23	1.01058E+00	1.03752E+00	1.02666E+00	2.59681
0.24	1.01152E+00	1.04090E+00	1.02905E+00	2.49556
0.25	1.01250E+00	1.04444E+00	1.03154E+00	2.40271
0.26	1.01352E+00	1.04813E+00	1.03414E+00	2.31729
0.27	1.01458E+00	1.05197E+00	1.03685E+00	2.23847
0.28	1.01568E+00	1.05596E+00	1.03966E+00	2.16555
0.29	1.01682E+00	1.06012E+00	1.04258E+00	2.09793
0.30	1.01800E+00	1.06443E+00	1.04561E+00	2.03507
0.31	1.01922E+00	1.06890E+00	1.04874E+00	1.97651
0.32	1.02048E+00	1.07353E+00	1.05199E+00	1.92185
0.33	1.02178E+00	1.07833E+00	1.05534E+00	1.87074
0.34	1.02312E+00	1.08329E+00	1.05881E+00	1.82288
0.35	1.02450E+00	1.08841E+00	1.06238E+00	1.77797
0.36	1.02592E+00	1.09370E+00	1.06607E+00	1.73578
0.37	1.02738E+00	1.09915E+00	1.06986E+00	1.69609
0.38	1.02888E+00	1.10478E+00	1.07377E+00	1.65870
0.39	1.03042E+00	1.11058E+00	1.07779E+00	1.62343

(continued)

Table A.1 (continued)

M	$\frac{T_0}{T}$	$\frac{P_0}{P}$	$\frac{\rho_0}{\rho}$	$\frac{A}{A^*}$
0.40	1.03200E+00	1.11655E+00	1.08193E+00	1.59014
0.41	1.03362E+00	1.12270E+00	1.08618E+00	1.55867
0.42	1.03528E+00	1.12902E+00	1.09055E+00	1.52890
0.43	1.03698E+00	1.13552E+00	1.09503E+00	1.50072
0.44	1.03872E+00	1.14221E+00	1.09963E+00	1.47401
0.45	1.04050E+00	1.14907E+00	1.10435E+00	1.44867
0.46	1.04232E+00	1.15612E+00	1.10918E+00	1.42463
0.47	1.04418E+00	1.16336E+00	1.11414E+00	1.40180
0.48	1.04608E+00	1.17078E+00	1.11921E+00	1.38010
0.49	1.04802E+00	1.17840E+00	1.12441E+00	1.35947
0.50	1.05000E+00	1.18621E+00	1.12973E+00	1.33984
0.51	1.05202E+00	1.19422E+00	1.13517E+00	1.32117
0.52	1.05408E+00	1.20242E+00	1.14073E+00	1.30339
0.53	1.05618E+00	1.21083E+00	1.14642E+00	1.28645
0.54	1.05832E+00	1.21944E+00	1.15224E+00	1.27032
0.55	1.06050E+00	1.22825E+00	1.15818E+00	1.25495
0.56	1.06272E+00	1.23727E+00	1.16425E+00	1.24029
0.57	1.06498E+00	1.24651E+00	1.17045E+00	1.22633
0.58	1.06728E+00	1.25596E+00	1.17678E+00	1.21301
0.59	1.06962E+00	1.26562E+00	1.18324E+00	1.20031
0.60	1.07200E+00	1.27550E+00	1.18984E+00	1.18820
0.61	1.07442E+00	1.28561E+00	1.19656E+00	1.17665
0.62	1.07688E+00	1.29594E+00	1.20342E+00	1.16565
0.63	1.07938E+00	1.30650E+00	1.21042E+00	1.15515
0.64	1.08192E+00	1.31729E+00	1.21755E+00	1.14515
0.65	1.08450E+00	1.32832E+00	1.22482E+00	1.13562
0.66	1.08712E+00	1.33959E+00	1.23224E+00	1.12654
0.67	1.08978E+00	1.35110E+00	1.23979E+00	1.11789
0.68	1.09248E+00	1.36285E+00	1.24748E+00	1.10965
0.69	1.09522E+00	1.37485E+00	1.25532E+00	1.10182
0.70	1.09800E+00	1.38710E+00	1.26330E+00	1.09437
0.71	1.10082E+00	1.39961E+00	1.27143E+00	1.08729
0.72	1.10368E+00	1.41238E+00	1.27970E+00	1.08057
0.73	1.10658E+00	1.42541E+00	1.28812E+00	1.07419
0.74	1.10952E+00	1.43871E+00	1.29670E+00	1.06814
0.75	1.11250E+00	1.45228E+00	1.30542E+00	1.06242
0.76	1.11552E+00	1.46612E+00	1.31430E+00	1.05700
0.77	1.11858E+00	1.48025E+00	1.32333E+00	1.05188
0.78	1.12168E+00	1.49466E+00	1.33252E+00	1.04705

(continued)

Table A.1 (continued)

M	$\frac{T_0}{T}$	$\frac{P_0}{P}$	$\frac{\rho_0}{\rho}$	$\frac{A}{A^*}$
0.79	1.12482E+00	1.50935E+00	1.34186E+00	1.04251
0.80	1.12800E+00	1.52434E+00	1.35137E+00	1.03823
0.81	1.13122E+00	1.53962E+00	1.36103E+00	1.03422
0.82	1.13448E+00	1.55521E+00	1.37086E+00	1.03046
0.83	1.13778E+00	1.57110E+00	1.38085E+00	1.02696
0.84	1.14112E+00	1.58730E+00	1.39100E+00	1.02370
0.85	1.14450E+00	1.60382E+00	1.40133E+00	1.02067
0.86	1.14792E+00	1.62066E+00	1.41182E+00	1.01787
0.87	1.15138E+00	1.63782E+00	1.42248E+00	1.01530
0.88	1.15488E+00	1.65531E+00	1.43332E+00	1.01294
0.89	1.15842E+00	1.67314E+00	1.44433E+00	1.01080
0.90	1.16200E+00	1.69130E+00	1.45551E+00	1.00886
0.91	1.16562E+00	1.70982E+00	1.46687E+00	1.00713
0.92	1.16928E+00	1.72868E+00	1.47841E+00	1.00560
0.93	1.17298E+00	1.74790E+00	1.49014E+00	1.00426
0.94	1.17672E+00	1.76749E+00	1.50204E+00	1.00311
0.95	1.18050E+00	1.78744E+00	1.51414E+00	1.00215
0.96	1.18432E+00	1.80776E+00	1.52642E+00	1.00136
0.97	1.18818E+00	1.82847E+00	1.53888E+00	1.00076
0.98	1.19208E+00	1.84956E+00	1.55154E+00	1.00034
0.99	1.19602E+00	1.87105E+00	1.56439E+00	1.00008
1.00	1.20000E+00	1.89293E+00	1.57744E+00	1.00000
1.01	1.20402E+00	1.91522E+00	1.59069E+00	1.00008
1.02	1.20808E+00	1.93792E+00	1.60413E+00	1.00033
1.03	1.21218E+00	1.96103E+00	1.61777E+00	1.00074
1.04	1.21632E+00	1.98457E+00	1.63162E+00	1.00131
1.05	1.22050E+00	2.00855E+00	1.64568E+00	1.00203
1.06	1.22472E+00	2.03296E+00	1.65994E+00	1.00291
1.07	1.22898E+00	2.05782E+00	1.67441E+00	1.00394
1.08	1.23328E+00	2.08313E+00	1.68910E+00	1.00512
1.09	1.23762E+00	2.10890E+00	1.70399E+00	1.00645
1.10	1.24200E+00	2.13514E+00	1.71911E+00	1.00793
1.11	1.24642E+00	2.16185E+00	1.73445E+00	1.00955
1.12	1.25088E+00	2.18905E+00	1.75000E+00	1.01131
1.13	1.25538E+00	2.21673E+00	1.76579E+00	1.01322
1.14	1.25992E+00	2.24492E+00	1.78179E+00	1.01527
1.15	1.26450E+00	2.27361E+00	1.79803E+00	1.01745
1.16	1.26912E+00	2.30282E+00	1.81450E+00	1.01978
1.17	1.27378E+00	2.33255E+00	1.83120E+00	1.02224

(continued)

Table A.1 (continued)

M	$\dfrac{T_0}{T}$	$\dfrac{P_0}{P}$	$\dfrac{\rho_0}{\rho}$	$\dfrac{A}{A^*}$
1.18	1.27848E+00	2.36281E+00	1.84814E+00	1.02484
1.19	1.28322E+00	2.39361E+00	1.86532E+00	1.02757
1.20	1.28800E+00	2.42497E+00	1.88274E+00	1.03044
1.21	1.29282E+00	2.45688E+00	1.90040E+00	1.03344
1.22	1.29768E+00	2.48935E+00	1.91831E+00	1.03657
1.23	1.30258E+00	2.52241E+00	1.93647E+00	1.03983
1.24	1.30752E+00	2.55605E+00	1.95488E+00	1.04323
1.25	1.31250E+00	2.59029E+00	1.97355E+00	1.04675
1.26	1.31752E+00	2.62513E+00	1.99248E+00	1.05041
1.27	1.32258E+00	2.66058E+00	2.01166E+00	1.05419
1.28	1.32768E+00	2.69666E+00	2.03111E+00	1.05810
1.29	1.33282E+00	2.73338E+00	2.05083E+00	1.06214
1.30	1.33800E+00	2.77074E+00	2.07081E+00	1.06630
1.31	1.34322E+00	2.80876E+00	2.09107E+00	1.07060
1.32	1.34848E+00	2.84745E+00	2.11160E+00	1.07502
1.33	1.35378E+00	2.88681E+00	2.13241E+00	1.07957
1.34	1.35912E+00	2.92686E+00	2.15350E+00	1.08424
1.35	1.36450E+00	2.96761E+00	2.17487E+00	1.08904
1.36	1.36992E+00	3.00908E+00	2.19653E+00	1.09396
1.37	1.37538E+00	3.05126E+00	2.21849E+00	1.09902
1.38	1.38088E+00	3.09418E+00	2.24073E+00	1.10419
1.39	1.38642E+00	3.13785E+00	2.26327E+00	1.10950
1.40	1.39200E+00	3.18227E+00	2.28612E+00	1.11493
1.41	1.39762E+00	3.22747E+00	2.30926E+00	1.12048
1.42	1.40328E+00	3.27345E+00	2.33271E+00	1.12616
1.43	1.40898E+00	3.32022E+00	2.35647E+00	1.13197
1.44	1.41472E+00	3.36780E+00	2.38054E+00	1.13790
1.45	1.42050E+00	3.41621E+00	2.40493E+00	1.14396
1.46	1.42632E+00	3.46545E+00	2.42964E+00	1.15015
1.47	1.43218E+00	3.51554E+00	2.45468E+00	1.15646
1.48	1.43808E+00	3.56649E+00	2.48003E+00	1.16290
1.49	1.44402E+00	3.61831E+00	2.50572E+00	1.16947
1.50	1.45000E+00	3.67103E+00	2.53175E+00	1.17617
1.51	1.45602E+00	3.72465E+00	2.55810E+00	1.18299
1.52	1.46208E+00	3.77919E+00	2.58481E+00	1.18994
1.53	1.46818E+00	3.83467E+00	2.61185E+00	1.19702
1.54	1.47432E+00	3.89109E+00	2.63924E+00	1.20423
1.55	1.48050E+00	3.94848E+00	2.66699E+00	1.21157
1.56	1.48672E+00	4.00684E+00	2.69509E+00	1.21904

(continued)

Table A.1 (continued)

M	$\frac{T_0}{T}$	$\frac{P_0}{P}$	$\frac{\rho_0}{\rho}$	$\frac{A}{A^*}$
1.57	1.49298E+00	4.06620E+00	2.72355E+00	1.22664
1.58	1.49928E+00	4.12657E+00	2.75237E+00	1.23438
1.59	1.50562E+00	4.18797E+00	2.78156E+00	1.24224
1.60	1.51200E+00	4.25041E+00	2.81112E+00	1.25024
1.61	1.51842E+00	4.31392E+00	2.84106E+00	1.25836
1.62	1.52488E+00	4.37849E+00	2.87137E+00	1.26663
1.63	1.53138E+00	4.44417E+00	2.90207E+00	1.27502
1.64	1.53792E+00	4.51095E+00	2.93315E+00	1.28355
1.65	1.54450E+00	4.57886E+00	2.96463E+00	1.29222
1.66	1.55112E+00	4.64792E+00	2.99649E+00	1.30102
1.67	1.55778E+00	4.71815E+00	3.02876E+00	1.30996
1.68	1.56448E+00	4.78955E+00	3.06143E+00	1.31904
1.69	1.57122E+00	4.86216E+00	3.09451E+00	1.32825
1.70	1.57800E+00	4.93599E+00	3.12801E+00	1.33761
1.71	1.58482E+00	5.01106E+00	3.16191E+00	1.34710
1.72	1.59168E+00	5.08739E+00	3.19624E+00	1.35674
1.73	1.59858E+00	5.16500E+00	3.23099E+00	1.36651
1.74	1.60552E+00	5.24391E+00	3.26617E+00	1.37643
1.75	1.61250E+00	5.32413E+00	3.30179E+00	1.38649
1.76	1.61952E+00	5.40570E+00	3.33784E+00	1.39670
1.77	1.62658E+00	5.48863E+00	3.37434E+00	1.40705
1.78	1.63368E+00	5.57294E+00	3.41128E+00	1.41755
1.79	1.64082E+00	5.65866E+00	3.44868E+00	1.42819
1.80	1.64800E+00	5.74580E+00	3.48653E+00	1.43898
1.81	1.65522E+00	5.83438E+00	3.52484E+00	1.44992
1.82	1.66248E+00	5.92444E+00	3.56362E+00	1.46101
1.83	1.66978E+00	6.01599E+00	3.60287E+00	1.47225
1.84	1.67712E+00	6.10906E+00	3.64259E+00	1.48365
1.85	1.68450E+00	6.20367E+00	3.68279E+00	1.49519
1.86	1.69192E+00	6.29984E+00	3.72348E+00	1.50689
1.87	1.69938E+00	6.39760E+00	3.76466E+00	1.51875
1.88	1.70688E+00	6.49696E+00	3.80634E+00	1.53076
1.89	1.71442E+00	6.59797E+00	3.84851E+00	1.54293
1.90	1.72200E+00	6.70064E+00	3.89119E+00	1.55526
1.91	1.72962E+00	6.80499E+00	3.93438E+00	1.56774
1.92	1.73728E+00	6.91106E+00	3.97809E+00	1.58039
1.93	1.74498E+00	7.01886E+00	4.02232E+00	1.59320
1.94	1.75272E+00	7.12843E+00	4.06707E+00	1.60617
1.95	1.76050E+00	7.23979E+00	4.11235E+00	1.61931
1.96	1.76832E+00	7.35297E+00	4.15817E+00	1.63261

(continued)

Table A.1 (continued)

M	$\frac{T_0}{T}$	$\frac{P_0}{P}$	$\frac{\rho_0}{\rho}$	$\frac{A}{A^*}$
1.97	1.77618E+00	7.46800E+00	4.20453E+00	1.64608
1.98	1.78408E+00	7.58490E+00	4.25144E+00	1.65972
1.99	1.79202E+00	7.70371E+00	4.29890E+00	1.67352
2.00	1.80000E+00	7.82445E+00	4.34692E+00	1.68750
2.02	1.81608E+00	8.07184E+00	4.44465E+00	1.71597
2.04	1.83232E+00	8.32731E+00	4.54468E+00	1.74514
2.06	1.84872E+00	8.59110E+00	4.64706E+00	1.77502
2.08	1.86528E+00	8.86348E+00	4.75182E+00	1.80561
2.10	1.88200E+00	9.14468E+00	4.85902E+00	1.83694
2.12	1.89888E+00	9.43499E+00	4.96871E+00	1.86902
2.14	1.91592E+00	9.73466E+00	5.08093E+00	1.90184
2.16	1.93312E+00	1.00440E+01	5.19574E+00	1.93544
2.18	1.95048E+00	1.03632E+01	5.31317E+00	1.96981
2.20	1.96800E+00	1.06927E+01	5.43329E+00	2.00497
2.22	1.98568E+00	1.10327E+01	5.55614E+00	2.04094
2.24	2.00352E+00	1.13836E+01	5.68178E+00	2.07773
2.26	2.02152E+00	1.17455E+01	5.81025E+00	2.11535
2.28	2.03968E+00	1.21190E+01	5.94162E+00	2.15381
2.30	2.05800E+00	1.25043E+01	6.07594E+00	2.19313
2.32	2.07648E+00	1.29017E+01	6.21326E+00	2.23332
2.34	2.09512E+00	1.33116E+01	6.35363E+00	2.27440
2.36	2.11392E+00	1.37344E+01	6.49713E+00	2.31638
2.38	2.13288E+00	1.41704E+01	6.64379E+00	2.35928
2.40	2.15200E+00	1.46200E+01	6.79369E+00	2.40310
2.42	2.17128E+00	1.50836E+01	6.94688E+00	2.44787
2.44	2.19072E+00	1.55616E+01	7.10341E+00	2.49360
2.46	2.21032E+00	1.60544E+01	7.26337E+00	2.54031
2.48	2.23008E+00	1.65623E+01	7.42679E+00	2.58801
2.50	2.25000E+00	1.70859E+01	7.59375E+00	2.63672
2.52	2.27008E+00	1.76256E+01	7.76431E+00	2.68645
2.54	2.29032E+00	1.81818E+01	7.93854E+00	2.73723
2.56	2.31072E+00	1.87549E+01	8.11649E+00	2.78906
2.58	2.33128E+00	1.93455E+01	8.29824E+00	2.84197
2.60	2.35200E+00	1.99540E+01	8.48386E+00	2.89598
2.62	2.37288E+00	2.05809E+01	8.67340E+00	2.95109
2.64	2.39392E+00	2.12268E+01	8.86695E+00	3.00733
2.66	2.41512E+00	2.18920E+01	9.06456E+00	3.06472
2.68	2.43648E+00	2.25772E+01	9.26632E+00	3.12327
2.70	2.45800E+00	2.32829E+01	9.47228E+00	3.18301

(continued)

Table A.1 (continued)

M	$\frac{T_0}{T}$	$\frac{P_0}{P}$	$\frac{\rho_0}{\rho}$	$\frac{A}{A^*}$
2.72	2.47968E+00	2.40096E+01	9.68254E+00	3.24395
2.74	2.50152E+00	2.47579E+01	9.89715E+00	3.30611
2.76	2.52352E+00	2.55284E+01	1.01162E+01	3.36952
2.78	2.54568E+00	2.63217E+01	1.03397E+01	3.43418
2.80	2.56800E+00	2.71383E+01	1.05679E+01	3.50012
2.82	2.59048E+00	2.79789E+01	1.08007E+01	3.56737
2.84	2.61312E+00	2.88441E+01	1.10382E+01	3.63593
2.86	2.63592E+00	2.97346E+01	1.12806E+01	3.70584
2.88	2.65888E+00	3.06511E+01	1.15278E+01	3.77711
2.90	2.68200E+00	3.15941E+01	1.17800E+01	3.84977
2.92	2.70528E+00	3.25644E+01	1.20373E+01	3.92383
2.94	2.72872E+00	3.35627E+01	1.22998E+01	3.99932
2.96	2.75232E+00	3.45897E+01	1.25675E+01	4.07625
2.98	2.77608E+00	3.56461E+01	1.28404E+01	4.15466
3.00	2.80000E+00	3.67327E+01	1.31188E+01	4.23457
3.10	2.92200E+00	4.26462E+01	1.45949E+01	4.65731
3.20	3.04800E+00	4.94370E+01	1.62195E+01	5.12096
3.30	3.17800E+00	5.72188E+01	1.80047E+01	5.62865
3.40	3.31200E+00	6.61175E+01	1.99630E+01	6.18370
3.50	3.45000E+00	7.62723E+01	2.21079E+01	6.78962
3.60	3.59200E+00	8.78369E+01	2.44535E+01	7.45011
3.70	3.73800E+00	1.00981E+02	2.70146E+01	8.16907
3.80	3.88800E+00	1.15889E+02	2.98068E+01	8.95059
3.90	4.04200E+00	1.32766E+02	3.28466E+01	9.79897
4.00	4.20000E+00	1.51835E+02	3.61512E+01	10.71875
4.10	4.36200E+00	1.73340E+02	3.97387E+01	11.71465
4.20	4.52800E+00	1.97548E+02	4.36281E+01	12.79164
4.30	4.69800E+00	2.24748E+02	4.78390E+01	13.95490
4.40	4.87200E+00	2.55256E+02	5.23924E+01	15.20987
4.50	5.05000E+00	2.89414E+02	5.73097E+01	16.56219
4.60	5.23200E+00	3.27595E+02	6.26137E+01	18.01779
4.70	5.41800E+00	3.70200E+02	6.83278E+01	19.58283
4.80	5.60800E+00	4.17665E+02	7.44766E+01	21.26371
4.90	5.80200E+00	4.70459E+02	8.10857E+01	23.06712
5.00	6.00000E+00	5.29090E+02	8.81816E+01	25.00000

Table A.2 Normal shock properties for $\gamma = 1.4$

M_1	M_2	$\frac{P_2}{P_1}$	$\frac{T_2}{T_1}$	$\frac{\rho_2}{\rho_1}$	$\frac{P_{0,2}}{P_{0,1}}$
1.00	1.00000E+00	1.00000E+00	1.00000E+00	1.00000E+00	1.00000E+00
1.02	9.80519E-01	1.04713E+00	1.01325E+00	1.03344E+00	9.99990E-01
1.04	9.62025E-01	1.09520E+00	1.02634E+00	1.06709E+00	9.99923E-01
1.06	9.44445E-01	1.14420E+00	1.03931E+00	1.10092E+00	9.99751E-01
1.08	9.27713E-01	1.19413E+00	1.05217E+00	1.13492E+00	9.99431E-01
1.10	9.11770E-01	1.24500E+00	1.06494E+00	1.16908E+00	9.98928E-01
1.12	8.96563E-01	1.29680E+00	1.07763E+00	1.20338E+00	9.98213E-01
1.14	8.82042E-01	1.34953E+00	1.09027E+00	1.23779E+00	9.97261E-01
1.16	8.68162E-01	1.40320E+00	1.10287E+00	1.27231E+00	9.96052E-01
1.18	8.54884E-01	1.45780E+00	1.11544E+00	1.30693E+00	9.94569E-01
1.20	8.42170E-01	1.51333E+00	1.12799E+00	1.34161E+00	9.92798E-01
1.22	8.29986E-01	1.56980E+00	1.14054E+00	1.37636E+00	9.90731E-01
1.24	8.18301E-01	1.62720E+00	1.15309E+00	1.41116E+00	9.88359E-01
1.26	8.07085E-01	1.68553E+00	1.16566E+00	1.44599E+00	9.85677E-01
1.28	7.96312E-01	1.74480E+00	1.17825E+00	1.48084E+00	9.82682E-01
1.30	7.85957E-01	1.80500E+00	1.19087E+00	1.51570E+00	9.79374E-01
1.32	7.75997E-01	1.86613E+00	1.20353E+00	1.55055E+00	9.75752E-01
1.34	7.66412E-01	1.92820E+00	1.21624E+00	1.58538E+00	9.71819E-01
1.36	7.57181E-01	1.99120E+00	1.22900E+00	1.62018E+00	9.67579E-01
1.38	7.48286E-01	2.05513E+00	1.24181E+00	1.65494E+00	9.63035E-01
1.40	7.39709E-01	2.12000E+00	1.25469E+00	1.68966E+00	9.58194E-01
1.42	7.31436E-01	2.18580E+00	1.26764E+00	1.72430E+00	9.53063E-01
1.44	7.23451E-01	2.25253E+00	1.28066E+00	1.75888E+00	9.47648E-01
1.46	7.15740E-01	2.32020E+00	1.29377E+00	1.79337E+00	9.41958E-01
1.48	7.08290E-01	2.38880E+00	1.30695E+00	1.82777E+00	9.36001E-01
1.50	7.01089E-01	2.45833E+00	1.32022E+00	1.86207E+00	9.29787E-01
1.52	6.94125E-01	2.52880E+00	1.33357E+00	1.89626E+00	9.23324E-01
1.54	6.87388E-01	2.60020E+00	1.34703E+00	1.93033E+00	9.16624E-01
1.56	6.80867E-01	2.67253E+00	1.36057E+00	1.96427E+00	9.09697E-01
1.58	6.74553E-01	2.74580E+00	1.37422E+00	1.99808E+00	9.02552E-01
1.60	6.68437E-01	2.82000E+00	1.38797E+00	2.03175E+00	8.95200E-01
1.62	6.62511E-01	2.89513E+00	1.40182E+00	2.06526E+00	8.87653E-01
1.64	6.56765E-01	2.97120E+00	1.41578E+00	2.09863E+00	8.79921E-01
1.66	6.51194E-01	3.04820E+00	1.42985E+00	2.13183E+00	8.72014E-01
1.68	6.45789E-01	3.12613E+00	1.44403E+00	2.16486E+00	8.63944E-01
1.70	6.40544E-01	3.20500E+00	1.45833E+00	2.19772E+00	8.55721E-01
1.72	6.35452E-01	3.28480E+00	1.47274E+00	2.23040E+00	8.47356E-01
1.74	6.30508E-01	3.36553E+00	1.48727E+00	2.26289E+00	8.38860E-01
1.76	6.25705E-01	3.44720E+00	1.50192E+00	2.29520E+00	8.30242E-01
1.78	6.21037E-01	3.52980E+00	1.51669E+00	2.32731E+00	8.21513E-01

(continued)

Table A.2 (continued)

M_1	M_2	$\dfrac{P_2}{P_1}$	$\dfrac{T_2}{T_1}$	$\dfrac{\rho_2}{\rho_1}$	$\dfrac{P_{0,2}}{P_{0,1}}$
1.80	6.16501E-01	3.61333E+00	1.53158E+00	2.35922E+00	8.12684E-01
1.82	6.12091E-01	3.69780E+00	1.54659E+00	2.39093E+00	8.03763E-01
1.84	6.07802E-01	3.78320E+00	1.56173E+00	2.42244E+00	7.94761E-01
1.86	6.03629E-01	3.86953E+00	1.57700E+00	2.45373E+00	7.85686E-01
1.88	5.99569E-01	3.95680E+00	1.59239E+00	2.48481E+00	7.76549E-01
1.90	5.95616E-01	4.04500E+00	1.60792E+00	2.51568E+00	7.67357E-01
1.92	5.91769E-01	4.13413E+00	1.62357E+00	2.54633E+00	7.58119E-01
1.94	5.88022E-01	4.22420E+00	1.63935E+00	2.57675E+00	7.48844E-01
1.96	5.84372E-01	4.31520E+00	1.65527E+00	2.60695E+00	7.39540E-01
1.98	5.80816E-01	4.40713E+00	1.67132E+00	2.63692E+00	7.30214E-01
2.00	5.77350E-01	4.50000E+00	1.68750E+00	2.66667E+00	7.20874E-01
2.02	5.73972E-01	4.59380E+00	1.70382E+00	2.69618E+00	7.11527E-01
2.04	5.70679E-01	4.68853E+00	1.72027E+00	2.72546E+00	7.02180E-01
2.06	5.67467E-01	4.78420E+00	1.73686E+00	2.75451E+00	6.92839E-01
2.08	5.64334E-01	4.88080E+00	1.75359E+00	2.78332E+00	6.83512E-01
2.10	5.61277E-01	4.97833E+00	1.77045E+00	2.81190E+00	6.74203E-01
2.12	5.58294E-01	5.07680E+00	1.78745E+00	2.84024E+00	6.64919E-01
2.14	5.55383E-01	5.17620E+00	1.80459E+00	2.86835E+00	6.55666E-01
2.16	5.52541E-01	5.27653E+00	1.82188E+00	2.89621E+00	6.46447E-01
2.18	5.49766E-01	5.37780E+00	1.83930E+00	2.92383E+00	6.37269E-01
2.20	5.47056E-01	5.48000E+00	1.85686E+00	2.95122E+00	6.28136E-01
2.22	5.44409E-01	5.58313E+00	1.87456E+00	2.97837E+00	6.19053E-01
2.24	5.41822E-01	5.68720E+00	1.89241E+00	3.00527E+00	6.10023E-01
2.26	5.39295E-01	5.79220E+00	1.91040E+00	3.03194E+00	6.01051E-01
2.28	5.36825E-01	5.89813E+00	1.92853E+00	3.05836E+00	5.92140E-01
2.30	5.34411E-01	6.00500E+00	1.94680E+00	3.08455E+00	5.83295E-01
2.32	5.32051E-01	6.11280E+00	1.96522E+00	3.11049E+00	5.74517E-01
2.34	5.29743E-01	6.22153E+00	1.98378E+00	3.13620E+00	5.65810E-01
2.36	5.27486E-01	6.33120E+00	2.00249E+00	3.16167E+00	5.57177E-01
2.38	5.25278E-01	6.44180E+00	2.02134E+00	3.18690E+00	5.48621E-01
2.40	5.23118E-01	6.55333E+00	2.04033E+00	3.21190E+00	5.40144E-01
2.42	5.21004E-01	6.66580E+00	2.05947E+00	3.23665E+00	5.31748E-01
2.44	5.18936E-01	6.77920E+00	2.07876E+00	3.26117E+00	5.23435E-01
2.46	5.16911E-01	6.89353E+00	2.09819E+00	3.28546E+00	5.15208E-01
2.48	5.14929E-01	7.00880E+00	2.11777E+00	3.30951E+00	5.07067E-01
2.50	5.12989E-01	7.12500E+00	2.13750E+00	3.33333E+00	4.99015E-01
2.52	5.11089E-01	7.24213E+00	2.15737E+00	3.35692E+00	4.91052E-01
2.54	5.09228E-01	7.36020E+00	2.17739E+00	3.38028E+00	4.83181E-01
2.56	5.07406E-01	7.47920E+00	2.19756E+00	3.40341E+00	4.75402E-01
2.58	5.05620E-01	7.59913E+00	2.21788E+00	3.42631E+00	4.67715E-01

(continued)

Table A.2 (continued)

M_1	M_2	$\frac{P_2}{P_1}$	$\frac{T_2}{T_1}$	$\frac{\rho_2}{\rho_1}$	$\frac{P_{0,2}}{P_{0,1}}$
2.60	5.03871E-01	7.72000E+00	2.23834E+00	3.44898E+00	4.60123E-01
2.62	5.02157E-01	7.84180E+00	2.25896E+00	3.47143E+00	4.52625E-01
2.64	5.00477E-01	7.96453E+00	2.27972E+00	3.49365E+00	4.45223E-01
2.66	4.98830E-01	8.08820E+00	2.30063E+00	3.51565E+00	4.37916E-01
2.68	4.97216E-01	8.21280E+00	2.32168E+00	3.53743E+00	4.30705E-01
2.70	4.95634E-01	8.33833E+00	2.34289E+00	3.55899E+00	4.23590E-01
2.72	4.94082E-01	8.46480E+00	2.36425E+00	3.58033E+00	4.16572E-01
2.74	4.92560E-01	8.59220E+00	2.38576E+00	3.60146E+00	4.09650E-01
2.76	4.91068E-01	8.72053E+00	2.40741E+00	3.62237E+00	4.02825E-01
2.78	4.89604E-01	8.84980E+00	2.42922E+00	3.64307E+00	3.96096E-01
2.80	4.88167E-01	8.98000E+00	2.45117E+00	3.66355E+00	3.89464E-01
2.82	4.86758E-01	9.11113E+00	2.47328E+00	3.68383E+00	3.82927E-01
2.84	4.85376E-01	9.24320E+00	2.49554E+00	3.70389E+00	3.76486E-01
2.86	4.84019E-01	9.37620E+00	2.51794E+00	3.72375E+00	3.70141E-01
2.88	4.82687E-01	9.51013E+00	2.54050E+00	3.74341E+00	3.63890E-01
2.90	4.81380E-01	9.64500E+00	2.56321E+00	3.76286E+00	3.57733E-01
2.92	4.80096E-01	9.78080E+00	2.58607E+00	3.78211E+00	3.51670E-01
2.94	4.78836E-01	9.91753E+00	2.60908E+00	3.80117E+00	3.45701E-01
2.96	4.77599E-01	1.00552E+01	2.63224E+00	3.82002E+00	3.39823E-01
2.98	4.76384E-01	1.01938E+01	2.65555E+00	3.83868E+00	3.34038E-01
3.00	4.75191E-01	1.03333E+01	2.67901E+00	3.85714E+00	3.28344E-01
3.02	4.74019E-01	1.04738E+01	2.70263E+00	3.87541E+00	3.22740E-01
3.04	4.72868E-01	1.06152E+01	2.72639E+00	3.89350E+00	3.17226E-01
3.06	4.71737E-01	1.07575E+01	2.75031E+00	3.91139E+00	3.11800E-01
3.08	4.70625E-01	1.09008E+01	2.77438E+00	3.92909E+00	3.06462E-01
3.10	4.69534E-01	1.10450E+01	2.79860E+00	3.94661E+00	3.01211E-01
3.12	4.68460E-01	1.11901E+01	2.82298E+00	3.96395E+00	2.96046E-01
3.14	4.67406E-01	1.13362E+01	2.84750E+00	3.98110E+00	2.90967E-01
3.16	4.66369E-01	1.14832E+01	2.87218E+00	3.99808E+00	2.85971E-01
3.18	4.65350E-01	1.16311E+01	2.89701E+00	4.01488E+00	2.81059E-01
3.20	4.64349E-01	1.17800E+01	2.92199E+00	4.03150E+00	2.76229E-01
3.22	4.63364E-01	1.19298E+01	2.94713E+00	4.04794E+00	2.71480E-01
3.24	4.62395E-01	1.20805E+01	2.97241E+00	4.06422E+00	2.66811E-01
3.26	4.61443E-01	1.22322E+01	2.99785E+00	4.08032E+00	2.62221E-01
3.28	4.60507E-01	1.23848E+01	3.02345E+00	4.09625E+00	2.57710E-01
3.30	4.59586E-01	1.25383E+01	3.04919E+00	4.11202E+00	2.53276E-01
3.32	4.58680E-01	1.26928E+01	3.07509E+00	4.12762E+00	2.48918E-01
3.34	4.57788E-01	1.28482E+01	3.10114E+00	4.14306E+00	2.44635E-01
3.36	4.56912E-01	1.30045E+01	3.12734E+00	4.15833E+00	2.40426E-01
3.38	4.56049E-01	1.31618E+01	3.15370E+00	4.17345E+00	2.36290E-01

(continued)

Table A.2 (continued)

M_1	M_2	$\frac{P_2}{P_1}$	$\frac{T_2}{T_1}$	$\frac{\rho_2}{\rho_1}$	$\frac{P_{0,2}}{P_{0,1}}$
3.40	4.55200E-01	1.33200E+01	3.18021E+00	4.18841E+00	2.32226E-01
3.42	4.54365E-01	1.34791E+01	3.20687E+00	4.20321E+00	2.28232E-01
3.44	4.53543E-01	1.36392E+01	3.23369E+00	4.21785E+00	2.24309E-01
3.46	4.52734E-01	1.38002E+01	3.26065E+00	4.23234E+00	2.20454E-01
3.48	4.51938E-01	1.39621E+01	3.28778E+00	4.24668E+00	2.16668E-01
3.50	4.51154E-01	1.41250E+01	3.31505E+00	4.26087E+00	2.12948E-01
3.52	4.50382E-01	1.42888E+01	3.34248E+00	4.27491E+00	2.09293E-01
3.54	4.49623E-01	1.44535E+01	3.37006E+00	4.28880E+00	2.05704E-01
3.56	4.48875E-01	1.46192E+01	3.39780E+00	4.30255E+00	2.02177E-01
3.58	4.48138E-01	1.47858E+01	3.42569E+00	4.31616E+00	1.98714E-01
3.60	4.47413E-01	1.49533E+01	3.45373E+00	4.32962E+00	1.95312E-01
3.62	4.46699E-01	1.51218E+01	3.48192E+00	4.34294E+00	1.91971E-01
3.64	4.45995E-01	1.52912E+01	3.51027E+00	4.35613E+00	1.88690E-01
3.66	4.45302E-01	1.54615E+01	3.53878E+00	4.36918E+00	1.85467E-01
3.68	4.44620E-01	1.56328E+01	3.56743E+00	4.38209E+00	1.82302E-01
3.70	4.43948E-01	1.58050E+01	3.59624E+00	4.39486E+00	1.79194E-01
3.72	4.43285E-01	1.59781E+01	3.62521E+00	4.40751E+00	1.76141E-01
3.74	4.42633E-01	1.61522E+01	3.65433E+00	4.42002E+00	1.73143E-01
3.76	4.41990E-01	1.63272E+01	3.68360E+00	4.43241E+00	1.70200E-01
3.78	4.41356E-01	1.65031E+01	3.71302E+00	4.44466E+00	1.67309E-01
3.80	4.40732E-01	1.66800E+01	3.74260E+00	4.45679E+00	1.64470E-01
3.82	4.40117E-01	1.68578E+01	3.77234E+00	4.46879E+00	1.61683E-01
3.84	4.39510E-01	1.70365E+01	3.80223E+00	4.48067E+00	1.58946E-01
3.86	4.38912E-01	1.72162E+01	3.83227E+00	4.49243E+00	1.56258E-01
3.88	4.38323E-01	1.73968E+01	3.86246E+00	4.50407E+00	1.53619E-01
3.90	4.37742E-01	1.75783E+01	3.89281E+00	4.51559E+00	1.51027E-01
3.92	4.37170E-01	1.77608E+01	3.92332E+00	4.52699E+00	1.48483E-01
3.94	4.36605E-01	1.79442E+01	3.95398E+00	4.53827E+00	1.45984E-01
3.96	4.36049E-01	1.81285E+01	3.98479E+00	4.54944E+00	1.43531E-01
3.98	4.35500E-01	1.83138E+01	4.01575E+00	4.56049E+00	1.41122E-01
4.00	4.34959E-01	1.85000E+01	4.04688E+00	4.57143E+00	1.38756E-01
4.02	4.34425E-01	1.86871E+01	4.07815E+00	4.58226E+00	1.36434E-01
4.04	4.33899E-01	1.88752E+01	4.10958E+00	4.59298E+00	1.34153E-01
4.06	4.33380E-01	1.90642E+01	4.14116E+00	4.60359E+00	1.31914E-01
4.08	4.32868E-01	1.92541E+01	4.17290E+00	4.61409E+00	1.29715E-01
4.10	4.32363E-01	1.94450E+01	4.20479E+00	4.62448E+00	1.27556E-01
4.12	4.31865E-01	1.96368E+01	4.23684E+00	4.63478E+00	1.25436E-01
4.14	4.31373E-01	1.98295E+01	4.26904E+00	4.64496E+00	1.23355E-01
4.16	4.30888E-01	2.00232E+01	4.30140E+00	4.65505E+00	1.21311E-01
4.18	4.30410E-01	2.02178E+01	4.33391E+00	4.66503E+00	1.19304E-01

(continued)

Table A.2 (continued)

M_1	M_2	$\dfrac{P_2}{P_1}$	$\dfrac{T_2}{T_1}$	$\dfrac{\rho_2}{\rho_1}$	$\dfrac{P_{0,2}}{P_{0,1}}$
4.20	4.29938E-01	2.04133E+01	4.36657E+00	4.67491E+00	1.17334E-01
4.22	4.29472E-01	2.06098E+01	4.39939E+00	4.68470E+00	1.15399E-01
4.24	4.29012E-01	2.08072E+01	4.43236E+00	4.69438E+00	1.13498E-01
4.26	4.28559E-01	2.10055E+01	4.46549E+00	4.70397E+00	1.11633E-01
4.28	4.28111E-01	2.12048E+01	4.49877E+00	4.71346E+00	1.09801E-01
4.30	4.27669E-01	2.14050E+01	4.53221E+00	4.72286E+00	1.08002E-01
4.32	4.27233E-01	2.16061E+01	4.56580E+00	4.73217E+00	1.06235E-01
4.34	4.26803E-01	2.18082E+01	4.59955E+00	4.74138E+00	1.04500E-01
4.36	4.26378E-01	2.20112E+01	4.63345E+00	4.75050E+00	1.02796E-01
4.38	4.25959E-01	2.22151E+01	4.66750E+00	4.75953E+00	1.01124E-01
4.40	4.25545E-01	2.24200E+01	4.70171E+00	4.76847E+00	9.94806E-02
4.42	4.25136E-01	2.26258E+01	4.73608E+00	4.77733E+00	9.78673E-02
4.44	4.24732E-01	2.28325E+01	4.77060E+00	4.78609E+00	9.62828E-02
4.46	4.24334E-01	2.30402E+01	4.80527E+00	4.79477E+00	9.47268E-02
4.48	4.23940E-01	2.32488E+01	4.84010E+00	4.80337E+00	9.31987E-02
4.50	4.23552E-01	2.34583E+01	4.87509E+00	4.81188E+00	9.16979E-02
4.52	4.23168E-01	2.36688E+01	4.91022E+00	4.82031E+00	9.02239E-02
4.54	4.22789E-01	2.38802E+01	4.94552E+00	4.82866E+00	8.87763E-02
4.56	4.22415E-01	2.40925E+01	4.98097E+00	4.83692E+00	8.73545E-02
4.58	4.22045E-01	2.43058E+01	5.01657E+00	4.84511E+00	8.59580E-02
4.60	4.21680E-01	2.45200E+01	5.05233E+00	4.85321E+00	8.45865E-02
4.62	4.21319E-01	2.47351E+01	5.08824E+00	4.86124E+00	8.32393E-02
4.64	4.20963E-01	2.49512E+01	5.12430E+00	4.86919E+00	8.19161E-02
4.66	4.20611E-01	2.51682E+01	5.16053E+00	4.87706E+00	8.06164E-02
4.68	4.20263E-01	2.53861E+01	5.19690E+00	4.88486E+00	7.93397E-02
4.70	4.19920E-01	2.56050E+01	5.23343E+00	4.89258E+00	7.80856E-02
4.72	4.19581E-01	2.58248E+01	5.27012E+00	4.90023E+00	7.68537E-02
4.74	4.19245E-01	2.60455E+01	5.30696E+00	4.90780E+00	7.56436E-02
4.76	4.18914E-01	2.62672E+01	5.34396E+00	4.91531E+00	7.44548E-02
4.78	4.18586E-01	2.64898E+01	5.38111E+00	4.92274E+00	7.32870E-02
4.80	4.18263E-01	2.67133E+01	5.41842E+00	4.93010E+00	7.21398E-02
4.82	4.17943E-01	2.69378E+01	5.45588E+00	4.93739E+00	7.10127E-02
4.84	4.17627E-01	2.71632E+01	5.49349E+00	4.94461E+00	6.99054E-02
4.86	4.17315E-01	2.73895E+01	5.53126E+00	4.95177E+00	6.88176E-02
4.88	4.17006E-01	2.76168E+01	5.56919E+00	4.95885E+00	6.77487E-02
4.90	4.16701E-01	2.78450E+01	5.60727E+00	4.96587E+00	6.66986E-02
4.92	4.16400E-01	2.80741E+01	5.64551E+00	4.97283E+00	6.56668E-02
4.94	4.16101E-01	2.83042E+01	5.68390E+00	4.97972E+00	6.46531E-02
4.96	4.15807E-01	2.85352E+01	5.72244E+00	4.98654E+00	6.36569E-02
4.98	4.15515E-01	2.87671E+01	5.76114E+00	4.99330E+00	6.26781E-02
5.00	4.15227E-01	2.90000E+01	5.80000E+00	5.00000E+00	6.17163E-02

Table A.3 Rayleigh flow properties for $\gamma = 1.4$

M	$\dfrac{P}{P^*}$	$\dfrac{T}{T^*}$	$\dfrac{\rho}{\rho^*}$	$\dfrac{P_0}{P_0^*}$	$\dfrac{T_0}{T_0^*}$
0.01	2.39966E+00	5.75839E-04	4.16725E+03	1.26779E+00	4.79875E-04
0.02	2.39866E+00	2.30142E-03	1.04225E+03	1.26752E+00	1.91800E-03
0.03	2.39698E+00	5.17096E-03	4.63546E+02	1.26708E+00	4.30991E-03
0.04	2.39464E+00	9.17485E-03	2.61000E+02	1.26646E+00	7.64816E-03
0.05	2.39163E+00	1.42997E-02	1.67250E+02	1.26567E+00	1.19224E-02
0.06	2.38796E+00	2.05286E-02	1.16324E+02	1.26470E+00	1.71194E-02
0.07	2.38365E+00	2.78407E-02	8.56173E+01	1.26356E+00	2.32233E-02
0.08	2.37869E+00	3.62122E-02	6.56875E+01	1.26226E+00	3.02154E-02
0.09	2.37309E+00	4.56156E-02	5.20237E+01	1.26078E+00	3.80746E-02
0.10	2.36686E+00	5.60204E-02	4.22500E+01	1.25915E+00	4.67771E-02
0.11	2.36002E+00	6.73934E-02	3.50186E+01	1.25735E+00	5.62971E-02
0.12	2.35257E+00	7.96982E-02	2.95185E+01	1.25539E+00	6.66064E-02
0.13	2.34453E+00	9.28962E-02	2.52382E+01	1.25329E+00	7.76751E-02
0.14	2.33590E+00	1.06946E-01	2.18418E+01	1.25103E+00	8.94712E-02
0.15	2.32671E+00	1.21805E-01	1.91019E+01	1.24863E+00	1.01961E-01
0.16	2.31696E+00	1.37429E-01	1.68594E+01	1.24608E+00	1.15110E-01
0.17	2.30667E+00	1.53769E-01	1.50009E+01	1.24340E+00	1.28882E-01
0.18	2.29586E+00	1.70779E-01	1.34434E+01	1.24059E+00	1.43238E-01
0.19	2.28454E+00	1.88410E-01	1.21253E+01	1.23765E+00	1.58142E-01
0.20	2.27273E+00	2.06612E-01	1.10000E+01	1.23460E+00	1.73554E-01
0.21	2.26044E+00	2.25333E-01	1.00316E+01	1.23142E+00	1.89434E-01
0.22	2.24770E+00	2.44523E-01	9.19215E+00	1.22814E+00	2.05742E-01
0.23	2.23451E+00	2.64132E-01	8.45983E+00	1.22475E+00	2.22439E-01
0.24	2.22091E+00	2.84108E-01	7.81713E+00	1.22126E+00	2.39484E-01
0.25	2.20690E+00	3.04400E-01	7.25000E+00	1.21767E+00	2.56837E-01
0.26	2.19250E+00	3.24957E-01	6.74704E+00	1.21400E+00	2.74459E-01
0.27	2.17774E+00	3.45732E-01	6.29893E+00	1.21025E+00	2.92311E-01
0.28	2.16263E+00	3.66674E-01	5.89796E+00	1.20642E+00	3.10353E-01
0.29	2.14719E+00	3.87737E-01	5.53775E+00	1.20251E+00	3.28549E-01
0.30	2.13144E+00	4.08873E-01	5.21296E+00	1.19855E+00	3.46860E-01
0.31	2.11539E+00	4.30037E-01	4.91909E+00	1.19452E+00	3.65252E-01
0.32	2.09908E+00	4.51187E-01	4.65234E+00	1.19045E+00	3.83689E-01
0.33	2.08250E+00	4.72279E-01	4.40947E+00	1.18632E+00	4.02138E-01
0.34	2.06569E+00	4.93273E-01	4.18772E+00	1.18215E+00	4.20565E-01
0.35	2.04866E+00	5.14131E-01	3.98469E+00	1.17795E+00	4.38940E-01
0.36	2.03142E+00	5.34816E-01	3.79835E+00	1.17371E+00	4.57232E-01
0.37	2.01400E+00	5.55292E-01	3.62692E+00	1.16945E+00	4.75413E-01
0.38	1.99641E+00	5.75526E-01	3.46884E+00	1.16517E+00	4.93456E-01
0.39	1.97866E+00	5.95488E-01	3.32276E+00	1.16088E+00	5.11336E-01

(continued)

Table A.3 (continued)

M	$\dfrac{P}{P^*}$	$\dfrac{T}{T^*}$	$\dfrac{\rho}{\rho^*}$	$\dfrac{P_0}{P_0^*}$	$\dfrac{T_0}{T_0^*}$
0.40	1.96078E+00	6.15148E-01	3.18750E+00	1.15658E+00	5.29027E-01
0.41	1.94278E+00	6.34479E-01	3.06202E+00	1.15227E+00	5.46508E-01
0.42	1.92468E+00	6.53456E-01	2.94539E+00	1.14796E+00	5.63758E-01
0.43	1.90649E+00	6.72055E-01	2.83680E+00	1.14366E+00	5.80756E-01
0.44	1.88822E+00	6.90255E-01	2.73554E+00	1.13936E+00	5.97485E-01
0.45	1.86989E+00	7.08037E-01	2.64095E+00	1.13508E+00	6.13927E-01
0.46	1.85151E+00	7.25383E-01	2.55246E+00	1.13082E+00	6.30068E-01
0.47	1.83310E+00	7.42278E-01	2.46956E+00	1.12659E+00	6.45893E-01
0.48	1.81466E+00	7.58707E-01	2.39178E+00	1.12238E+00	6.61390E-01
0.49	1.79622E+00	7.74659E-01	2.31872E+00	1.11820E+00	6.76549E-01
0.50	1.77778E+00	7.90123E-01	2.25000E+00	1.11405E+00	6.91358E-01
0.51	1.75935E+00	8.05091E-01	2.18528E+00	1.10995E+00	7.05810E-01
0.52	1.74095E+00	8.19554E-01	2.12426E+00	1.10588E+00	7.19897E-01
0.53	1.72258E+00	8.33508E-01	2.06666E+00	1.10186E+00	7.33612E-01
0.54	1.70425E+00	8.46948E-01	2.01223E+00	1.09789E+00	7.46952E-01
0.55	1.68599E+00	8.59870E-01	1.96074E+00	1.09397E+00	7.59910E-01
0.56	1.66778E+00	8.72274E-01	1.91199E+00	1.09011E+00	7.72486E-01
0.57	1.64964E+00	8.84158E-01	1.86578E+00	1.08630E+00	7.84675E-01
0.58	1.63159E+00	8.95523E-01	1.82194E+00	1.08256E+00	7.96478E-01
0.59	1.61362E+00	9.06371E-01	1.78031E+00	1.07887E+00	8.07894E-01
0.60	1.59574E+00	9.16704E-01	1.74074E+00	1.07525E+00	8.18923E-01
0.61	1.57797E+00	9.26527E-01	1.70310E+00	1.07170E+00	8.29566E-01
0.62	1.56031E+00	9.35843E-01	1.66727E+00	1.06822E+00	8.39825E-01
0.63	1.54275E+00	9.44657E-01	1.63314E+00	1.06481E+00	8.49703E-01
0.64	1.52532E+00	9.52976E-01	1.60059E+00	1.06147E+00	8.59203E-01
0.65	1.50801E+00	9.60806E-01	1.56953E+00	1.05821E+00	8.68329E-01
0.66	1.49083E+00	9.68155E-01	1.53987E+00	1.05503E+00	8.77084E-01
0.67	1.47379E+00	9.75030E-01	1.51153E+00	1.05193E+00	8.85473E-01
0.68	1.45688E+00	9.81439E-01	1.48443E+00	1.04890E+00	8.93502E-01
0.69	1.44011E+00	9.87391E-01	1.45850E+00	1.04596E+00	9.01175E-01
0.70	1.42349E+00	9.92895E-01	1.43367E+00	1.04310E+00	9.08499E-01
0.71	1.40701E+00	9.97961E-01	1.40989E+00	1.04033E+00	9.15479E-01
0.72	1.39069E+00	1.00260E+00	1.38709E+00	1.03764E+00	9.22122E-01
0.73	1.37452E+00	1.00682E+00	1.36522E+00	1.03504E+00	9.28435E-01
0.74	1.35851E+00	1.01062E+00	1.34423E+00	1.03253E+00	9.34423E-01
0.75	1.34266E+00	1.01403E+00	1.32407E+00	1.03010E+00	9.40095E-01
0.76	1.32696E+00	1.01706E+00	1.30471E+00	1.02777E+00	9.45456E-01
0.77	1.31143E+00	1.01970E+00	1.28609E+00	1.02552E+00	9.50515E-01
0.78	1.29606E+00	1.02198E+00	1.26819E+00	1.02337E+00	9.55279E-01

(continued)

Table A.3 (continued)

M	$\dfrac{P}{P^*}$	$\dfrac{T}{T^*}$	$\dfrac{\rho}{\rho^*}$	$\dfrac{P_0}{P_0^*}$	$\dfrac{T_0}{T_0^*}$
0.79	1.28086E+00	1.02390E+00	1.25096E+00	1.02131E+00	9.59754E-01
0.80	1.26582E+00	1.02548E+00	1.23438E+00	1.01934E+00	9.63948E-01
0.81	1.25095E+00	1.02672E+00	1.21840E+00	1.01747E+00	9.67869E-01
0.82	1.23625E+00	1.02763E+00	1.20300E+00	1.01569E+00	9.71524E-01
0.83	1.22171E+00	1.02823E+00	1.18816E+00	1.01400E+00	9.74921E-01
0.84	1.20734E+00	1.02853E+00	1.17385E+00	1.01241E+00	9.78066E-01
0.85	1.19314E+00	1.02854E+00	1.16003E+00	1.01091E+00	9.80968E-01
0.86	1.17911E+00	1.02826E+00	1.14670E+00	1.00951E+00	9.83633E-01
0.87	1.16524E+00	1.02771E+00	1.13382E+00	1.00820E+00	9.86069E-01
0.88	1.15154E+00	1.02689E+00	1.12138E+00	1.00699E+00	9.88283E-01
0.89	1.13801E+00	1.02583E+00	1.10936E+00	1.00587E+00	9.90282E-01
0.90	1.12465E+00	1.02452E+00	1.09774E+00	1.00486E+00	9.92073E-01
0.91	1.11145E+00	1.02297E+00	1.08649E+00	1.00393E+00	9.93663E-01
0.92	1.09842E+00	1.02120E+00	1.07561E+00	1.00311E+00	9.95058E-01
0.93	1.08555E+00	1.01922E+00	1.06508E+00	1.00238E+00	9.96266E-01
0.94	1.07285E+00	1.01702E+00	1.05489E+00	1.00175E+00	9.97293E-01
0.95	1.06030E+00	1.01463E+00	1.04501E+00	1.00122E+00	9.98145E-01
0.96	1.04793E+00	1.01205E+00	1.03545E+00	1.00078E+00	9.98828E-01
0.97	1.03571E+00	1.00929E+00	1.02617E+00	1.00044E+00	9.99350E-01
0.98	1.02365E+00	1.00636E+00	1.01718E+00	1.00019E+00	9.99715E-01
0.99	1.01174E+00	1.00326E+00	1.00846E+00	1.00005E+00	9.99930E-01
1.00	1.00000E+00	1.00000E+00	1.00000E+00	1.00000E+00	1.00000E+00
1.01	9.88411E-01	9.96593E-01	9.91790E-01	1.00005E+00	9.99931E-01
1.02	9.76976E-01	9.93043E-01	9.83820E-01	1.00019E+00	9.99730E-01
1.03	9.65694E-01	9.89358E-01	9.76082E-01	1.00044E+00	9.99400E-01
1.04	9.54563E-01	9.85543E-01	9.68565E-01	1.00078E+00	9.98947E-01
1.05	9.43582E-01	9.81607E-01	9.61262E-01	1.00122E+00	9.98376E-01
1.06	9.32749E-01	9.77555E-01	9.54165E-01	1.00175E+00	9.97692E-01
1.07	9.22063E-01	9.73393E-01	9.47266E-01	1.00238E+00	9.96901E-01
1.08	9.11522E-01	9.69129E-01	9.40558E-01	1.00311E+00	9.96006E-01
1.09	9.01124E-01	9.64767E-01	9.34033E-01	1.00394E+00	9.95012E-01
1.10	8.90869E-01	9.60313E-01	9.27686E-01	1.00486E+00	9.93924E-01
1.11	8.80753E-01	9.55773E-01	9.21509E-01	1.00588E+00	9.92745E-01
1.12	8.70777E-01	9.51151E-01	9.15497E-01	1.00699E+00	9.91480E-01
1.13	8.60937E-01	9.46455E-01	9.09644E-01	1.00821E+00	9.90133E-01
1.14	8.51233E-01	9.41687E-01	9.03945E-01	1.00952E+00	9.88708E-01
1.15	8.41662E-01	9.36853E-01	8.98393E-01	1.01093E+00	9.87209E-01
1.16	8.32224E-01	9.31958E-01	8.92985E-01	1.01243E+00	9.85638E-01
1.17	8.22915E-01	9.27005E-01	8.87714E-01	1.01403E+00	9.84001E-01

(continued)

Table A.3 (continued)

M	$\dfrac{P}{P^*}$	$\dfrac{T}{T^*}$	$\dfrac{\rho}{\rho^*}$	$\dfrac{P_0}{P_0^*}$	$\dfrac{T_0}{T_0^*}$
1.18	8.13736E-01	9.22000E-01	8.82577E-01	1.01573E+00	9.82299E-01
1.19	8.04683E-01	9.16946E-01	8.77569E-01	1.01752E+00	9.80536E-01
1.20	7.95756E-01	9.11848E-01	8.72685E-01	1.01942E+00	9.78717E-01
1.21	7.86952E-01	9.06708E-01	8.67922E-01	1.02140E+00	9.76842E-01
1.22	7.78271E-01	9.01532E-01	8.63276E-01	1.02349E+00	9.74916E-01
1.23	7.69709E-01	8.96321E-01	8.58743E-01	1.02567E+00	9.72942E-01
1.24	7.61267E-01	8.91081E-01	8.54318E-01	1.02795E+00	9.70922E-01
1.25	7.52941E-01	8.85813E-01	8.50000E-01	1.03033E+00	9.68858E-01
1.26	7.44731E-01	8.80522E-01	8.45784E-01	1.03280E+00	9.66754E-01
1.27	7.36635E-01	8.75209E-01	8.41667E-01	1.03537E+00	9.64612E-01
1.28	7.28651E-01	8.69878E-01	8.37646E-01	1.03803E+00	9.62433E-01
1.29	7.20777E-01	8.64532E-01	8.33719E-01	1.04080E+00	9.60222E-01
1.30	7.13012E-01	8.59174E-01	8.29882E-01	1.04366E+00	9.57979E-01
1.31	7.05355E-01	8.53805E-01	8.26132E-01	1.04662E+00	9.55706E-01
1.32	6.97804E-01	8.48428E-01	8.22467E-01	1.04968E+00	9.53407E-01
1.33	6.90357E-01	8.43046E-01	8.18885E-01	1.05283E+00	9.51082E-01
1.34	6.83013E-01	8.37661E-01	8.15382E-01	1.05608E+00	9.48734E-01
1.35	6.75771E-01	8.32274E-01	8.11957E-01	1.05943E+00	9.46365E-01
1.36	6.68628E-01	8.26888E-01	8.08607E-01	1.06288E+00	9.43976E-01
1.37	6.61584E-01	8.21505E-01	8.05331E-01	1.06642E+00	9.41569E-01
1.38	6.54636E-01	8.16127E-01	8.02125E-01	1.07007E+00	9.39145E-01
1.39	6.47784E-01	8.10755E-01	7.98988E-01	1.07381E+00	9.36706E-01
1.40	6.41026E-01	8.05391E-01	7.95918E-01	1.07765E+00	9.34254E-01
1.41	6.34360E-01	8.00037E-01	7.92914E-01	1.08159E+00	9.31790E-01
1.42	6.27786E-01	7.94694E-01	7.89972E-01	1.08563E+00	9.29315E-01
1.43	6.21301E-01	7.89363E-01	7.87092E-01	1.08977E+00	9.26830E-01
1.44	6.14905E-01	7.84046E-01	7.84272E-01	1.09401E+00	9.24338E-01
1.45	6.08596E-01	7.78744E-01	7.81510E-01	1.09835E+00	9.21838E-01
1.46	6.02373E-01	7.73459E-01	7.78805E-01	1.10278E+00	9.19333E-01
1.47	5.96235E-01	7.68191E-01	7.76154E-01	1.10732E+00	9.16823E-01
1.48	5.90179E-01	7.62942E-01	7.73557E-01	1.11196E+00	9.14310E-01
1.49	5.84206E-01	7.57713E-01	7.71013E-01	1.11670E+00	9.11794E-01
1.50	5.78313E-01	7.52504E-01	7.68519E-01	1.12155E+00	9.09276E-01
1.51	5.72500E-01	7.47317E-01	7.66074E-01	1.12649E+00	9.06757E-01
1.52	5.66765E-01	7.42152E-01	7.63677E-01	1.13153E+00	9.04238E-01
1.53	5.61107E-01	7.37011E-01	7.61328E-01	1.13668E+00	9.01721E-01
1.54	5.55525E-01	7.31894E-01	7.59023E-01	1.14193E+00	8.99205E-01
1.55	5.50017E-01	7.26802E-01	7.56764E-01	1.14729E+00	8.96692E-01
1.56	5.44583E-01	7.21735E-01	7.54547E-01	1.15274E+00	8.94181E-01
1.57	5.39222E-01	7.16694E-01	7.52373E-01	1.15830E+00	8.91675E-01

(continued)

Table A.3 (continued)

M	$\dfrac{P}{P^*}$	$\dfrac{T}{T^*}$	$\dfrac{\rho}{\rho^*}$	$\dfrac{P_0}{P_0^*}$	$\dfrac{T_0}{T_0^*}$
1.58	5.33931E-01	7.11680E-01	7.50240E-01	1.16397E+00	8.89173E-01
1.59	5.28711E-01	7.06694E-01	7.48147E-01	1.16974E+00	8.86677E-01
1.60	5.23560E-01	7.01735E-01	7.46094E-01	1.17561E+00	8.84186E-01
1.61	5.18477E-01	6.96805E-01	7.44078E-01	1.18159E+00	8.81702E-01
1.62	5.13461E-01	6.91903E-01	7.42100E-01	1.18768E+00	8.79225E-01
1.63	5.08511E-01	6.87031E-01	7.40158E-01	1.19387E+00	8.76754E-01
1.64	5.03626E-01	6.82188E-01	7.38251E-01	1.20017E+00	8.74292E-01
1.65	4.98805E-01	6.77375E-01	7.36379E-01	1.20657E+00	8.71839E-01
1.66	4.94047E-01	6.72593E-01	7.34541E-01	1.21309E+00	8.69394E-01
1.67	4.89351E-01	6.67841E-01	7.32735E-01	1.21971E+00	8.66958E-01
1.68	4.84715E-01	6.63120E-01	7.30962E-01	1.22644E+00	8.64531E-01
1.69	4.80140E-01	6.58430E-01	7.29220E-01	1.23328E+00	8.62115E-01
1.70	4.75624E-01	6.53771E-01	7.27509E-01	1.24024E+00	8.59709E-01
1.71	4.71167E-01	6.49144E-01	7.25827E-01	1.24730E+00	8.57314E-01
1.72	4.66766E-01	6.44549E-01	7.24175E-01	1.25447E+00	8.54929E-01
1.73	4.62422E-01	6.39985E-01	7.22552E-01	1.26175E+00	8.52556E-01
1.74	4.58134E-01	6.35454E-01	7.20956E-01	1.26915E+00	8.50195E-01
1.75	4.53901E-01	6.30954E-01	7.19388E-01	1.27666E+00	8.47845E-01
1.76	4.49721E-01	6.26487E-01	7.17846E-01	1.28428E+00	8.45507E-01
1.77	4.45595E-01	6.22052E-01	7.16330E-01	1.29202E+00	8.43181E-01
1.78	4.41521E-01	6.17649E-01	7.14840E-01	1.29987E+00	8.40868E-01
1.79	4.37498E-01	6.13279E-01	7.13375E-01	1.30784E+00	8.38567E-01
1.80	4.33526E-01	6.08941E-01	7.11934E-01	1.31592E+00	8.36279E-01
1.81	4.29604E-01	6.04636E-01	7.10517E-01	1.32413E+00	8.34004E-01
1.82	4.25731E-01	6.00363E-01	7.09123E-01	1.33244E+00	8.31743E-01
1.83	4.21907E-01	5.96122E-01	7.07752E-01	1.34088E+00	8.29494E-01
1.84	4.18130E-01	5.91914E-01	7.06404E-01	1.34943E+00	8.27259E-01
1.85	4.14400E-01	5.87738E-01	7.05077E-01	1.35811E+00	8.25037E-01
1.86	4.10717E-01	5.83595E-01	7.03771E-01	1.36690E+00	8.22829E-01
1.87	4.07079E-01	5.79483E-01	7.02486E-01	1.37582E+00	8.20635E-01
1.88	4.03486E-01	5.75404E-01	7.01222E-01	1.38486E+00	8.18455E-01
1.89	3.99937E-01	5.71357E-01	6.99978E-01	1.39402E+00	8.16288E-01
1.90	3.96432E-01	5.67342E-01	6.98753E-01	1.40330E+00	8.14136E-01
1.91	3.92970E-01	5.63359E-01	6.97548E-01	1.41271E+00	8.11997E-01
1.92	3.89550E-01	5.59407E-01	6.96361E-01	1.42224E+00	8.09873E-01
1.93	3.86171E-01	5.55488E-01	6.95193E-01	1.43190E+00	8.07762E-01
1.94	3.82834E-01	5.51599E-01	6.94043E-01	1.44168E+00	8.05666E-01
1.95	3.79537E-01	5.47743E-01	6.92910E-01	1.45159E+00	8.03584E-01
1.96	3.76279E-01	5.43917E-01	6.91795E-01	1.46164E+00	8.01517E-01
1.97	3.73061E-01	5.40123E-01	6.90697E-01	1.47180E+00	7.99463E-01

(continued)

Table A.3 (continued)

M	$\frac{P}{P^*}$	$\frac{T}{T^*}$	$\frac{\rho}{\rho^*}$	$\frac{P_0}{P_0^*}$	$\frac{T_0}{T_0^*}$
1.98	3.69882E-01	5.36360E-01	6.89615E-01	1.48210E+00	7.97424E-01
1.99	3.66740E-01	5.32627E-01	6.88550E-01	1.49253E+00	7.95399E-01
2.00	3.63636E-01	5.28926E-01	6.87500E-01	1.50310E+00	7.93388E-01
2.02	3.57539E-01	5.21614E-01	6.85448E-01	1.52462E+00	7.89410E-01
2.04	3.51584E-01	5.14422E-01	6.83455E-01	1.54668E+00	7.85488E-01
2.06	3.45770E-01	5.07350E-01	6.81520E-01	1.56928E+00	7.81624E-01
2.08	3.40090E-01	5.00396E-01	6.79641E-01	1.59244E+00	7.77816E-01
2.10	3.34541E-01	4.93558E-01	6.77816E-01	1.61616E+00	7.74064E-01
2.12	3.29121E-01	4.86835E-01	6.76041E-01	1.64045E+00	7.70368E-01
2.14	3.23824E-01	4.80225E-01	6.74317E-01	1.66531E+00	7.66727E-01
2.16	3.18647E-01	4.73727E-01	6.72639E-01	1.69076E+00	7.63142E-01
2.18	3.13588E-01	4.67338E-01	6.71008E-01	1.71680E+00	7.59611E-01
2.20	3.08642E-01	4.61058E-01	6.69421E-01	1.74345E+00	7.56135E-01
2.22	3.03807E-01	4.54884E-01	6.67877E-01	1.77070E+00	7.52712E-01
2.24	2.99079E-01	4.48815E-01	6.66374E-01	1.79858E+00	7.49342E-01
2.26	2.94455E-01	4.42849E-01	6.64911E-01	1.82708E+00	7.46024E-01
2.28	2.89934E-01	4.36985E-01	6.63486E-01	1.85623E+00	7.42758E-01
2.30	2.85510E-01	4.31220E-01	6.62098E-01	1.88602E+00	7.39543E-01
2.32	2.81183E-01	4.25554E-01	6.60746E-01	1.91647E+00	7.36379E-01
2.34	2.76949E-01	4.19984E-01	6.59428E-01	1.94759E+00	7.33264E-01
2.36	2.72807E-01	4.14509E-01	6.58144E-01	1.97939E+00	7.30199E-01
2.38	2.68752E-01	4.09127E-01	6.56892E-01	2.01187E+00	7.27182E-01
2.40	2.64784E-01	4.03836E-01	6.55671E-01	2.04505E+00	7.24213E-01
2.42	2.60899E-01	3.98635E-01	6.54481E-01	2.07895E+00	7.21291E-01
2.44	2.57096E-01	3.93523E-01	6.53319E-01	2.11356E+00	7.18415E-01
2.46	2.53372E-01	3.88497E-01	6.52186E-01	2.14891E+00	7.15585E-01
2.48	2.49725E-01	3.83556E-01	6.51080E-01	2.18499E+00	7.12800E-01
2.50	2.46154E-01	3.78698E-01	6.50000E-01	2.22183E+00	7.10059E-01
2.52	2.42656E-01	3.73923E-01	6.48946E-01	2.25944E+00	7.07362E-01
2.54	2.39229E-01	3.69228E-01	6.47917E-01	2.29782E+00	7.04708E-01
2.56	2.35871E-01	3.64611E-01	6.46912E-01	2.33699E+00	7.02096E-01
2.58	2.32582E-01	3.60073E-01	6.45930E-01	2.37696E+00	6.99525E-01
2.60	2.29358E-01	3.55610E-01	6.44970E-01	2.41774E+00	6.96995E-01
2.62	2.26198E-01	3.51222E-01	6.44033E-01	2.45935E+00	6.94506E-01
2.64	2.23101E-01	3.46907E-01	6.43117E-01	2.50179E+00	6.92055E-01
2.66	2.20066E-01	3.42663E-01	6.42221E-01	2.54509E+00	6.89644E-01
2.68	2.17089E-01	3.38490E-01	6.41346E-01	2.58925E+00	6.87271E-01
2.70	2.14171E-01	3.34387E-01	6.40489E-01	2.63429E+00	6.84935E-01
2.72	2.11309E-01	3.30350E-01	6.39652E-01	2.68021E+00	6.82636E-01

(continued)

Table A.3 (continued)

M	$\frac{P}{P^*}$	$\frac{T}{T^*}$	$\frac{\rho}{\rho^*}$	$\frac{P_0}{P_0^*}$	$\frac{T_0}{T_0^*}$
2.74	2.08503E-01	3.26381E-01	6.38833E-01	2.72704E+00	6.80374E-01
2.76	2.05750E-01	3.22476E-01	6.38031E-01	2.77478E+00	6.78146E-01
2.78	2.03050E-01	3.18636E-01	6.37247E-01	2.82346E+00	6.75954E-01
2.80	2.00401E-01	3.14858E-01	6.36480E-01	2.87308E+00	6.73796E-01
2.82	1.97802E-01	3.11142E-01	6.35728E-01	2.92366E+00	6.71672E-01
2.84	1.95251E-01	3.07486E-01	6.34993E-01	2.97521E+00	6.69582E-01
2.86	1.92749E-01	3.03889E-01	6.34273E-01	3.02775E+00	6.67523E-01
2.88	1.90293E-01	3.00351E-01	6.33568E-01	3.08129E+00	6.65497E-01
2.90	1.87882E-01	2.96869E-01	6.32878E-01	3.13585E+00	6.63502E-01
2.92	1.85515E-01	2.93443E-01	6.32201E-01	3.19145E+00	6.61538E-01
2.94	1.83192E-01	2.90072E-01	6.31539E-01	3.24809E+00	6.59604E-01
2.96	1.80910E-01	2.86754E-01	6.30889E-01	3.30579E+00	6.57700E-01
2.98	1.78670E-01	2.83490E-01	6.30253E-01	3.36457E+00	6.55825E-01
3.00	1.76471E-01	2.80277E-01	6.29630E-01	3.42445E+00	6.53979E-01
3.10	1.66044E-01	2.64954E-01	6.26691E-01	3.74084E+00	6.45162E-01
3.20	1.56495E-01	2.50783E-01	6.24023E-01	4.08712E+00	6.36989E-01
3.30	1.47729E-01	2.37661E-01	6.21595E-01	4.46549E+00	6.29405E-01
3.40	1.39665E-01	2.25492E-01	6.19377E-01	4.87830E+00	6.22359E-01
3.50	1.32231E-01	2.14193E-01	6.17347E-01	5.32804E+00	6.15805E-01
3.60	1.25366E-01	2.03686E-01	6.15484E-01	5.81730E+00	6.09701E-01
3.70	1.19012E-01	1.93904E-01	6.13769E-01	6.34884E+00	6.04010E-01
3.80	1.13122E-01	1.84783E-01	6.12188E-01	6.92557E+00	5.98698E-01
3.90	1.07652E-01	1.76269E-01	6.10728E-01	7.55050E+00	5.93732E-01
4.00	1.02564E-01	1.68310E-01	6.09375E-01	8.22685E+00	5.89086E-01
4.10	9.78234E-02	1.60862E-01	6.08120E-01	8.95794E+00	5.84733E-01
4.20	9.33998E-02	1.53883E-01	6.06954E-01	9.74729E+00	5.80651E-01
4.30	8.92658E-02	1.47335E-01	6.05868E-01	1.05985E+01	5.76818E-01
4.40	8.53971E-02	1.41186E-01	6.04855E-01	1.15155E+01	5.73215E-01
4.50	8.17717E-02	1.35404E-01	6.03909E-01	1.25023E+01	5.69825E-01
4.60	7.83699E-02	1.29961E-01	6.03025E-01	1.35629E+01	5.66632E-01
4.70	7.51738E-02	1.24833E-01	6.02196E-01	1.47017E+01	5.63621E-01
4.80	7.21674E-02	1.19995E-01	6.01418E-01	1.59234E+01	5.60779E-01
4.90	6.93361E-02	1.15428E-01	6.00687E-01	1.72325E+01	5.58094E-01
5.00	6.66667E-02	1.11111E-01	6.00000E-01	1.86339E+01	5.55556E-01

Table A.4 Fanno flow properties for $\gamma = 1.4$

M	$\dfrac{P}{P^*}$	$\dfrac{T}{T^*}$	$\dfrac{\rho}{\rho^*}$	$\dfrac{P_0}{P_0^*}$	$\dfrac{fL^*}{D_h}$
0.01	1.09543E+02	1.19998E+00	9.12880E+01	5.78738E+01	7.13440E+03
0.02	5.47701E+01	1.19990E+00	4.56454E+01	2.89421E+01	1.77845E+03
0.03	3.65116E+01	1.19978E+00	3.04318E+01	1.93005E+01	7.87081E+02
0.04	2.73817E+01	1.19962E+00	2.28254E+01	1.44815E+01	4.40352E+02
0.05	2.19034E+01	1.19940E+00	1.82620E+01	1.15914E+01	2.80020E+02
0.06	1.82508E+01	1.19914E+00	1.52200E+01	9.66591E+00	1.93031E+02
0.07	1.56416E+01	1.19883E+00	1.30474E+01	8.29153E+00	1.40655E+02
0.08	1.36843E+01	1.19847E+00	1.14182E+01	7.26161E+00	1.06718E+02
0.09	1.21618E+01	1.19806E+00	1.01512E+01	6.46134E+00	8.34961E+01
0.10	1.09435E+01	1.19760E+00	9.13783E+00	5.82183E+00	6.69216E+01
0.11	9.94656E+00	1.19710E+00	8.30886E+00	5.29923E+00	5.46879E+01
0.12	9.11559E+00	1.19655E+00	7.61820E+00	4.86432E+00	4.54080E+01
0.13	8.41230E+00	1.19596E+00	7.03394E+00	4.49686E+00	3.82070E+01
0.14	7.80932E+00	1.19531E+00	6.53327E+00	4.18240E+00	3.25113E+01
0.15	7.28659E+00	1.19462E+00	6.09948E+00	3.91034E+00	2.79320E+01
0.16	6.82907E+00	1.19389E+00	5.72003E+00	3.67274E+00	2.41978E+01
0.17	6.42525E+00	1.19310E+00	5.38533E+00	3.46351E+00	2.11152E+01
0.18	6.06618E+00	1.19227E+00	5.08791E+00	3.27793E+00	1.85427E+01
0.19	5.74480E+00	1.19140E+00	4.82190E+00	3.11226E+00	1.63752E+01
0.20	5.45545E+00	1.19048E+00	4.58258E+00	2.96352E+00	1.45333E+01
0.21	5.19355E+00	1.18951E+00	4.36613E+00	2.82929E+00	1.29560E+01
0.22	4.95537E+00	1.18850E+00	4.16945E+00	2.70760E+00	1.15961E+01
0.23	4.73781E+00	1.18744E+00	3.98994E+00	2.59681E+00	1.04161E+01
0.24	4.53829E+00	1.18633E+00	3.82548E+00	2.49556E+00	9.38648E+00
0.25	4.35465E+00	1.18519E+00	3.67423E+00	2.40271E+00	8.48341E+00
0.26	4.18505E+00	1.18399E+00	3.53470E+00	2.31729E+00	7.68757E+00
0.27	4.02795E+00	1.18276E+00	3.40556E+00	2.23847E+00	6.98317E+00
0.28	3.88199E+00	1.18147E+00	3.28571E+00	2.16555E+00	6.35721E+00
0.29	3.74602E+00	1.18015E+00	3.17419E+00	2.09793E+00	5.79891E+00
0.30	3.61906E+00	1.17878E+00	3.07017E+00	2.03507E+00	5.29925E+00
0.31	3.50022E+00	1.17737E+00	2.97291E+00	1.97651E+00	4.85066E+00
0.32	3.38874E+00	1.17592E+00	2.88179E+00	1.92185E+00	4.44674E+00
0.33	3.28396E+00	1.17442E+00	2.79624E+00	1.87074E+00	4.08205E+00
0.34	3.18529E+00	1.17288E+00	2.71577E+00	1.82288E+00	3.75195E+00
0.35	3.09219E+00	1.17130E+00	2.63996E+00	1.77797E+00	3.45245E+00
0.36	3.00422E+00	1.16968E+00	2.56841E+00	1.73578E+00	3.18012E+00
0.37	2.92094E+00	1.16802E+00	2.50077E+00	1.69609E+00	2.93198E+00
0.38	2.84200E+00	1.16632E+00	2.43673E+00	1.65870E+00	2.70545E+00
0.39	2.76706E+00	1.16457E+00	2.37603E+00	1.62343E+00	2.49828E+00

(continued)

Table A.4 (continued)

M	$\dfrac{P}{P^*}$	$\dfrac{T}{T^*}$	$\dfrac{\rho}{\rho^*}$	$\dfrac{P_0}{P_0^*}$	$\dfrac{fL^*}{D_h}$
0.40	2.69582E+00	1.16279E+00	2.31840E+00	1.59014E+00	2.30849E+00
0.41	2.62801E+00	1.16097E+00	2.26363E+00	1.55867E+00	2.13436E+00
0.42	2.56338E+00	1.15911E+00	2.21151E+00	1.52890E+00	1.97437E+00
0.43	2.50171E+00	1.15721E+00	2.16185E+00	1.50072E+00	1.82715E+00
0.44	2.44280E+00	1.15527E+00	2.11449E+00	1.47401E+00	1.69152E+00
0.45	2.38648E+00	1.15329E+00	2.06927E+00	1.44867E+00	1.56643E+00
0.46	2.33256E+00	1.15128E+00	2.02606E+00	1.42463E+00	1.45091E+00
0.47	2.28089E+00	1.14923E+00	1.98472E+00	1.40180E+00	1.34413E+00
0.48	2.23135E+00	1.14714E+00	1.94514E+00	1.38010E+00	1.24534E+00
0.49	2.18378E+00	1.14502E+00	1.90721E+00	1.35947E+00	1.15385E+00
0.50	2.13809E+00	1.14286E+00	1.87083E+00	1.33984E+00	1.06906E+00
0.51	2.09415E+00	1.14066E+00	1.83591E+00	1.32117E+00	9.90414E-01
0.52	2.05187E+00	1.13843E+00	1.80237E+00	1.30339E+00	9.17418E-01
0.53	2.01116E+00	1.13617E+00	1.77012E+00	1.28645E+00	8.49624E-01
0.54	1.97192E+00	1.13387E+00	1.73910E+00	1.27032E+00	7.86625E-01
0.55	1.93407E+00	1.13154E+00	1.70924E+00	1.25495E+00	7.28053E-01
0.56	1.89755E+00	1.12918E+00	1.68047E+00	1.24029E+00	6.73571E-01
0.57	1.86228E+00	1.12678E+00	1.65274E+00	1.22633E+00	6.22874E-01
0.58	1.82820E+00	1.12435E+00	1.62600E+00	1.21301E+00	5.75683E-01
0.59	1.79525E+00	1.12189E+00	1.60019E+00	1.20031E+00	5.31743E-01
0.60	1.76336E+00	1.11940E+00	1.57527E+00	1.18820E+00	4.90822E-01
0.61	1.73250E+00	1.11688E+00	1.55120E+00	1.17665E+00	4.52705E-01
0.62	1.70261E+00	1.11433E+00	1.52792E+00	1.16565E+00	4.17197E-01
0.63	1.67364E+00	1.11175E+00	1.50541E+00	1.15515E+00	3.84116E-01
0.64	1.64556E+00	1.10914E+00	1.48363E+00	1.14515E+00	3.53299E-01
0.65	1.61831E+00	1.10650E+00	1.46255E+00	1.13562E+00	3.24591E-01
0.66	1.59187E+00	1.10383E+00	1.44213E+00	1.12654E+00	2.97853E-01
0.67	1.56620E+00	1.10114E+00	1.42234E+00	1.11789E+00	2.72955E-01
0.68	1.54126E+00	1.09842E+00	1.40316E+00	1.10965E+00	2.49775E-01
0.69	1.51702E+00	1.09567E+00	1.38456E+00	1.10182E+00	2.28204E-01
0.70	1.49345E+00	1.09290E+00	1.36651E+00	1.09437E+00	2.08139E-01
0.71	1.47053E+00	1.09010E+00	1.34899E+00	1.08729E+00	1.89483E-01
0.72	1.44823E+00	1.08727E+00	1.33198E+00	1.08057E+00	1.72149E-01
0.73	1.42652E+00	1.08442E+00	1.31546E+00	1.07419E+00	1.56054E-01
0.74	1.40537E+00	1.08155E+00	1.29941E+00	1.06814E+00	1.41122E-01
0.75	1.38478E+00	1.07865E+00	1.28380E+00	1.06242E+00	1.27282E-01
0.76	1.36470E+00	1.07573E+00	1.26863E+00	1.05700E+00	1.14468E-01
0.77	1.34514E+00	1.07279E+00	1.25387E+00	1.05188E+00	1.02617E-01
0.78	1.32605E+00	1.06982E+00	1.23951E+00	1.04705E+00	9.16722E-02

(continued)

Table A.4 (continued)

M	$\dfrac{P}{P^*}$	$\dfrac{T}{T^*}$	$\dfrac{\rho}{\rho^*}$	$\dfrac{P_0}{P_0^*}$	$\dfrac{fL^*}{D_h}$
0.79	1.30744E+00	1.06684E+00	1.22553E+00	1.04251E+00	8.15800E-02
0.80	1.28928E+00	1.06383E+00	1.21192E+00	1.03823E+00	7.22900E-02
0.81	1.27155E+00	1.06080E+00	1.19867E+00	1.03422E+00	6.37552E-02
0.82	1.25423E+00	1.05775E+00	1.18575E+00	1.03046E+00	5.59317E-02
0.83	1.23732E+00	1.05469E+00	1.17317E+00	1.02696E+00	4.87783E-02
0.84	1.22080E+00	1.05160E+00	1.16090E+00	1.02370E+00	4.22564E-02
0.85	1.20466E+00	1.04849E+00	1.14894E+00	1.02067E+00	3.63300E-02
0.86	1.18888E+00	1.04537E+00	1.13728E+00	1.01787E+00	3.09651E-02
0.87	1.17344E+00	1.04223E+00	1.12590E+00	1.01530E+00	2.61300E-02
0.88	1.15835E+00	1.03907E+00	1.11480E+00	1.01294E+00	2.17945E-02
0.89	1.14358E+00	1.03589E+00	1.10396E+00	1.01080E+00	1.79308E-02
0.90	1.12913E+00	1.03270E+00	1.09338E+00	1.00886E+00	1.45124E-02
0.91	1.11499E+00	1.02950E+00	1.08304E+00	1.00713E+00	1.15144E-02
0.92	1.10114E+00	1.02627E+00	1.07295E+00	1.00560E+00	8.91334E-03
0.93	1.08758E+00	1.02304E+00	1.06309E+00	1.00426E+00	6.68729E-03
0.94	1.07430E+00	1.01978E+00	1.05346E+00	1.00311E+00	4.81545E-03
0.95	1.06129E+00	1.01652E+00	1.04404E+00	1.00215E+00	3.27822E-03
0.96	1.04854E+00	1.01324E+00	1.03484E+00	1.00136E+00	2.05714E-03
0.97	1.03604E+00	1.00995E+00	1.02584E+00	1.00076E+00	1.13479E-03
0.98	1.02379E+00	1.00664E+00	1.01704E+00	1.00034E+00	4.94695E-04
0.99	1.01178E+00	1.00333E+00	1.00842E+00	1.00008E+00	1.21328E-04
1.00	1.00000E+00	1.00000E+00	1.00000E+00	1.00000E+00	0.00000E+00
1.01	9.88445E-01	9.96661E-01	9.91756E-01	1.00008E+00	1.16830E-04
1.02	9.77108E-01	9.93312E-01	9.83687E-01	1.00033E+00	4.58691E-04
1.03	9.65984E-01	9.89952E-01	9.75789E-01	1.00074E+00	1.01317E-03
1.04	9.55066E-01	9.86582E-01	9.68055E-01	1.00131E+00	1.76850E-03
1.05	9.44349E-01	9.83204E-01	9.60481E-01	1.00203E+00	2.71358E-03
1.06	9.33827E-01	9.79816E-01	9.53064E-01	1.00291E+00	3.83785E-03
1.07	9.23495E-01	9.76419E-01	9.45797E-01	1.00394E+00	5.13135E-03
1.08	9.13347E-01	9.73015E-01	9.38678E-01	1.00512E+00	6.58460E-03
1.09	9.03380E-01	9.69603E-01	9.31701E-01	1.00645E+00	8.18865E-03
1.10	8.93588E-01	9.66184E-01	9.24863E-01	1.00793E+00	9.93500E-03
1.11	8.83966E-01	9.62757E-01	9.18160E-01	1.00955E+00	1.18156E-02
1.12	8.74510E-01	9.59325E-01	9.11589E-01	1.01131E+00	1.38227E-02
1.13	8.65216E-01	9.55886E-01	9.05146E-01	1.01322E+00	1.59492E-02
1.14	8.56080E-01	9.52441E-01	8.98827E-01	1.01527E+00	1.81881E-02
1.15	8.47097E-01	9.48992E-01	8.92629E-01	1.01745E+00	2.05329E-02
1.16	8.38265E-01	9.45537E-01	8.86549E-01	1.01978E+00	2.29773E-02
1.17	8.29579E-01	9.42078E-01	8.80584E-01	1.02224E+00	2.55156E-02

(continued)

Table A.4 (continued)

M	$\dfrac{P}{P^*}$	$\dfrac{T}{T^*}$	$\dfrac{\rho}{\rho^*}$	$\dfrac{P_0}{P_0^*}$	$\dfrac{fL^*}{D_h}$
1.18	8.21035E-01	9.38615E-01	8.74731E-01	1.02484E+00	2.81419E-02
1.19	8.12630E-01	9.35148E-01	8.68986E-01	1.02757E+00	3.08511E-02
1.20	8.04362E-01	9.31677E-01	8.63348E-01	1.03044E+00	3.36381E-02
1.21	7.96226E-01	9.28203E-01	8.57814E-01	1.03344E+00	3.64980E-02
1.22	7.88219E-01	9.24727E-01	8.52380E-01	1.03657E+00	3.94262E-02
1.23	7.80339E-01	9.21249E-01	8.47045E-01	1.03983E+00	4.24184E-02
1.24	7.72582E-01	9.17768E-01	8.41806E-01	1.04323E+00	4.54705E-02
1.25	7.64946E-01	9.14286E-01	8.36660E-01	1.04675E+00	4.85785E-02
1.26	7.57428E-01	9.10802E-01	8.31606E-01	1.05041E+00	5.17387E-02
1.27	7.50025E-01	9.07318E-01	8.26640E-01	1.05419E+00	5.49474E-02
1.28	7.42735E-01	9.03832E-01	8.21762E-01	1.05810E+00	5.82014E-02
1.29	7.35555E-01	9.00347E-01	8.16969E-01	1.06214E+00	6.14973E-02
1.30	7.28483E-01	8.96861E-01	8.12258E-01	1.06630E+00	6.48321E-02
1.31	7.21516E-01	8.93376E-01	8.07629E-01	1.07060E+00	6.82028E-02
1.32	7.14652E-01	8.89891E-01	8.03078E-01	1.07502E+00	7.16067E-02
1.33	7.07889E-01	8.86407E-01	7.98605E-01	1.07957E+00	7.50411E-02
1.34	7.01224E-01	8.82924E-01	7.94207E-01	1.08424E+00	7.85035E-02
1.35	6.94656E-01	8.79443E-01	7.89882E-01	1.08904E+00	8.19915E-02
1.36	6.88183E-01	8.75964E-01	7.85630E-01	1.09396E+00	8.55027E-02
1.37	6.81803E-01	8.72486E-01	7.81448E-01	1.09902E+00	8.90350E-02
1.38	6.75513E-01	8.69011E-01	7.77335E-01	1.10419E+00	9.25863E-02
1.39	6.69312E-01	8.65539E-01	7.73289E-01	1.10950E+00	9.61547E-02
1.40	6.63198E-01	8.62069E-01	7.69309E-01	1.11493E+00	9.97382E-02
1.41	6.57169E-01	8.58602E-01	7.65394E-01	1.12048E+00	1.03335E-01
1.42	6.51224E-01	8.55139E-01	7.61541E-01	1.12616E+00	1.06943E-01
1.43	6.45360E-01	8.51680E-01	7.57750E-01	1.13197E+00	1.10562E-01
1.44	6.39577E-01	8.48224E-01	7.54019E-01	1.13790E+00	1.14189E-01
1.45	6.33873E-01	8.44773E-01	7.50347E-01	1.14396E+00	1.17823E-01
1.46	6.28245E-01	8.41326E-01	7.46732E-01	1.15015E+00	1.21462E-01
1.47	6.22694E-01	8.37884E-01	7.43175E-01	1.15646E+00	1.25106E-01
1.48	6.17216E-01	8.34446E-01	7.39672E-01	1.16290E+00	1.28753E-01
1.49	6.11812E-01	8.31013E-01	7.36224E-01	1.16947E+00	1.32401E-01
1.50	6.06478E-01	8.27586E-01	7.32828E-01	1.17617E+00	1.36050E-01
1.51	6.01215E-01	8.24165E-01	7.29485E-01	1.18299E+00	1.39699E-01
1.52	5.96021E-01	8.20749E-01	7.26192E-01	1.18994E+00	1.43346E-01
1.53	5.90894E-01	8.17338E-01	7.22949E-01	1.19702E+00	1.46990E-01
1.54	5.85833E-01	8.13935E-01	7.19755E-01	1.20423E+00	1.50631E-01
1.55	5.80838E-01	8.10537E-01	7.16608E-01	1.21157E+00	1.54268E-01
1.56	5.75906E-01	8.07146E-01	7.13509E-01	1.21904E+00	1.57899E-01

(continued)

Table A.4 (continued)

M	$\frac{P}{P^*}$	$\frac{T}{T^*}$	$\frac{\rho}{\rho^*}$	$\frac{P_0}{P_0^*}$	$\frac{fL^*}{D_h}$
1.57	5.71037E-01	8.03762E-01	7.10455E-01	1.22664E+00	1.61525E-01
1.58	5.66229E-01	8.00384E-01	7.07447E-01	1.23438E+00	1.65143E-01
1.59	5.61482E-01	7.97014E-01	7.04482E-01	1.24224E+00	1.68754E-01
1.60	5.56794E-01	7.93651E-01	7.01561E-01	1.25024E+00	1.72357E-01
1.61	5.52165E-01	7.90295E-01	6.98682E-01	1.25836E+00	1.75951E-01
1.62	5.47593E-01	7.86947E-01	6.95844E-01	1.26663E+00	1.79535E-01
1.63	5.43077E-01	7.83607E-01	6.93048E-01	1.27502E+00	1.83109E-01
1.64	5.38617E-01	7.80275E-01	6.90291E-01	1.28355E+00	1.86673E-01
1.65	5.34211E-01	7.76950E-01	6.87574E-01	1.29222E+00	1.90226E-01
1.66	5.29858E-01	7.73635E-01	6.84895E-01	1.30102E+00	1.93766E-01
1.67	5.25559E-01	7.70327E-01	6.82254E-01	1.30996E+00	1.97295E-01
1.68	5.21310E-01	7.67028E-01	6.79650E-01	1.31904E+00	2.00811E-01
1.69	5.17113E-01	7.63738E-01	6.77082E-01	1.32825E+00	2.04314E-01
1.70	5.12966E-01	7.60456E-01	6.74550E-01	1.33761E+00	2.07803E-01
1.71	5.08867E-01	7.57184E-01	6.72053E-01	1.34710E+00	2.11279E-01
1.72	5.04817E-01	7.53920E-01	6.69590E-01	1.35674E+00	2.14740E-01
1.73	5.00815E-01	7.50666E-01	6.67161E-01	1.36651E+00	2.18187E-01
1.74	4.96859E-01	7.47421E-01	6.64765E-01	1.37643E+00	2.21620E-01
1.75	4.92950E-01	7.44186E-01	6.62401E-01	1.38649E+00	2.25037E-01
1.76	4.89086E-01	7.40960E-01	6.60070E-01	1.39670E+00	2.28438E-01
1.77	4.85266E-01	7.37744E-01	6.57770E-01	1.40705E+00	2.31824E-01
1.78	4.81490E-01	7.34538E-01	6.55500E-01	1.41755E+00	2.35195E-01
1.79	4.77757E-01	7.31342E-01	6.53261E-01	1.42819E+00	2.38549E-01
1.80	4.74067E-01	7.28155E-01	6.51052E-01	1.43898E+00	2.41886E-01
1.81	4.70418E-01	7.24979E-01	6.48871E-01	1.44992E+00	2.45208E-01
1.82	4.66811E-01	7.21813E-01	6.46720E-01	1.46101E+00	2.48512E-01
1.83	4.63244E-01	7.18658E-01	6.44596E-01	1.47225E+00	2.51800E-01
1.84	4.59717E-01	7.15512E-01	6.42501E-01	1.48365E+00	2.55070E-01
1.85	4.56230E-01	7.12378E-01	6.40432E-01	1.49519E+00	2.58324E-01
1.86	4.52781E-01	7.09253E-01	6.38390E-01	1.50689E+00	2.61560E-01
1.87	4.49370E-01	7.06140E-01	6.36375E-01	1.51875E+00	2.64778E-01
1.88	4.45996E-01	7.03037E-01	6.34385E-01	1.53076E+00	2.67979E-01
1.89	4.42660E-01	6.99945E-01	6.32421E-01	1.54293E+00	2.71163E-01
1.90	4.39360E-01	6.96864E-01	6.30482E-01	1.55526E+00	2.74328E-01
1.91	4.36096E-01	6.93794E-01	6.28567E-01	1.56774E+00	2.77476E-01
1.92	4.32867E-01	6.90735E-01	6.26676E-01	1.58039E+00	2.80606E-01
1.93	4.29673E-01	6.87687E-01	6.24809E-01	1.59320E+00	2.83718E-01
1.94	4.26513E-01	6.84650E-01	6.22965E-01	1.60617E+00	2.86812E-01
1.95	4.23387E-01	6.81625E-01	6.21145E-01	1.61931E+00	2.89888E-01

(continued)

Table A.4 (continued)

M	$\dfrac{P}{P^*}$	$\dfrac{T}{T^*}$	$\dfrac{\rho}{\rho^*}$	$\dfrac{P_0}{P_0^*}$	$\dfrac{fL^*}{D_h}$
1.96	4.20295E-01	6.78610E-01	6.19347E-01	1.63261E+00	2.92946E-01
1.97	4.17235E-01	6.75607E-01	6.17571E-01	1.64608E+00	2.95986E-01
1.98	4.14208E-01	6.72616E-01	6.15817E-01	1.65972E+00	2.99008E-01
1.99	4.11212E-01	6.69635E-01	6.14084E-01	1.67352E+00	3.02011E-01
2.00	4.08248E-01	6.66667E-01	6.12372E-01	1.68750E+00	3.04997E-01
2.02	4.02413E-01	6.60764E-01	6.09012E-01	1.71597E+00	3.10913E-01
2.04	3.96698E-01	6.54907E-01	6.05731E-01	1.74514E+00	3.16756E-01
2.06	3.91100E-01	6.49098E-01	6.02529E-01	1.77502E+00	3.22526E-01
2.08	3.85616E-01	6.43335E-01	5.99402E-01	1.80561E+00	3.28225E-01
2.10	3.80243E-01	6.37620E-01	5.96348E-01	1.83694E+00	3.33851E-01
2.12	3.74978E-01	6.31951E-01	5.93365E-01	1.86902E+00	3.39405E-01
2.14	3.69818E-01	6.26331E-01	5.90452E-01	1.90184E+00	3.44887E-01
2.16	3.64760E-01	6.20758E-01	5.87604E-01	1.93544E+00	3.50299E-01
2.18	3.59802E-01	6.15233E-01	5.84822E-01	1.96981E+00	3.55639E-01
2.20	3.54940E-01	6.09756E-01	5.82102E-01	2.00497E+00	3.60910E-01
2.22	3.50173E-01	6.04327E-01	5.79443E-01	2.04094E+00	3.66111E-01
2.24	3.45498E-01	5.98946E-01	5.76844E-01	2.07773E+00	3.71243E-01
2.26	3.40913E-01	5.93613E-01	5.74302E-01	2.11535E+00	3.76306E-01
2.28	3.36415E-01	5.88328E-01	5.71815E-01	2.15381E+00	3.81302E-01
2.30	3.32002E-01	5.83090E-01	5.69383E-01	2.19313E+00	3.86230E-01
2.32	3.27672E-01	5.77901E-01	5.67003E-01	2.23332E+00	3.91092E-01
2.34	3.23423E-01	5.72760E-01	5.64674E-01	2.27440E+00	3.95888E-01
2.36	3.19253E-01	5.67666E-01	5.62395E-01	2.31638E+00	4.00619E-01
2.38	3.15160E-01	5.62620E-01	5.60165E-01	2.35928E+00	4.05286E-01
2.40	3.11142E-01	5.57621E-01	5.57981E-01	2.40310E+00	4.09889E-01
2.42	3.07197E-01	5.52669E-01	5.55843E-01	2.44787E+00	4.14429E-01
2.44	3.03324E-01	5.47765E-01	5.53749E-01	2.49360E+00	4.18907E-01
2.46	2.99521E-01	5.42908E-01	5.51699E-01	2.54031E+00	4.23324E-01
2.48	2.95787E-01	5.38097E-01	5.49690E-01	2.58801E+00	4.27680E-01
2.50	2.92119E-01	5.33333E-01	5.47723E-01	2.63672E+00	4.31977E-01
2.52	2.88516E-01	5.28616E-01	5.45795E-01	2.68645E+00	4.36214E-01
2.54	2.84976E-01	5.23944E-01	5.43906E-01	2.73723E+00	4.40393E-01
2.56	2.81499E-01	5.19319E-01	5.42055E-01	2.78906E+00	4.44514E-01
2.58	2.78083E-01	5.14739E-01	5.40240E-01	2.84197E+00	4.48579E-01
2.60	2.74725E-01	5.10204E-01	5.38462E-01	2.89598E+00	4.52588E-01
2.62	2.71426E-01	5.05715E-01	5.36718E-01	2.95109E+00	4.56541E-01
2.64	2.68183E-01	5.01270E-01	5.35008E-01	3.00733E+00	4.60441E-01
2.66	2.64996E-01	4.96870E-01	5.33331E-01	3.06472E+00	4.64286E-01
2.68	2.61863E-01	4.92514E-01	5.31687E-01	3.12327E+00	4.68079E-01

(continued)

Table A.4 (continued)

M	$\dfrac{P}{P^*}$	$\dfrac{T}{T^*}$	$\dfrac{\rho}{\rho^*}$	$\dfrac{P_0}{P_0^*}$	$\dfrac{fL^*}{D_h}$
2.70	2.58783E-01	4.88202E-01	5.30074E-01	3.18301E+00	4.71819E-01
2.72	2.55755E-01	4.83933E-01	5.28492E-01	3.24395E+00	4.75508E-01
2.74	2.52777E-01	4.79708E-01	5.26940E-01	3.30611E+00	4.79146E-01
2.76	2.49849E-01	4.75526E-01	5.25416E-01	3.36952E+00	4.82735E-01
2.78	2.46970E-01	4.71387E-01	5.23922E-01	3.43418E+00	4.86274E-01
2.80	2.44138E-01	4.67290E-01	5.22455E-01	3.50012E+00	4.89765E-01
2.82	2.41352E-01	4.63235E-01	5.21015E-01	3.56737E+00	4.93208E-01
2.84	2.38612E-01	4.59221E-01	5.19602E-01	3.63593E+00	4.96604E-01
2.86	2.35917E-01	4.55249E-01	5.18214E-01	3.70584E+00	4.99953E-01
2.88	2.33265E-01	4.51318E-01	5.16852E-01	3.77711E+00	5.03257E-01
2.90	2.30655E-01	4.47427E-01	5.15514E-01	3.84977E+00	5.06516E-01
2.92	2.28088E-01	4.43577E-01	5.14201E-01	3.92383E+00	5.09731E-01
2.94	2.25561E-01	4.39767E-01	5.12910E-01	3.99932E+00	5.12902E-01
2.96	2.23074E-01	4.35996E-01	5.11643E-01	4.07625E+00	5.16030E-01
2.98	2.20627E-01	4.32264E-01	5.10398E-01	4.15466E+00	5.19115E-01
3.00	2.18218E-01	4.28571E-01	5.09175E-01	4.23457E+00	5.22159E-01
3.10	2.06723E-01	4.10678E-01	5.03371E-01	4.65731E+00	5.36777E-01
3.20	1.96080E-01	3.93701E-01	4.98043E-01	5.12096E+00	5.50444E-01
3.30	1.86209E-01	3.77596E-01	4.93142E-01	5.62865E+00	5.63232E-01
3.40	1.77038E-01	3.62319E-01	4.88625E-01	6.18370E+00	5.75207E-01
3.50	1.68505E-01	3.47826E-01	4.84452E-01	6.78962E+00	5.86429E-01
3.60	1.60554E-01	3.34076E-01	4.80590E-01	7.45011E+00	5.96955E-01
3.70	1.53133E-01	3.21027E-01	4.77010E-01	8.16907E+00	6.06836E-01
3.80	1.46199E-01	3.08642E-01	4.73684E-01	8.95059E+00	6.16119E-01
3.90	1.39710E-01	2.96883E-01	4.70590E-01	9.79897E+00	6.24849E-01
4.00	1.33631E-01	2.85714E-01	4.67707E-01	1.07188E+01	6.33065E-01
4.10	1.27928E-01	2.75103E-01	4.65016E-01	1.17147E+01	6.40804E-01
4.20	1.22571E-01	2.65018E-01	4.62502E-01	1.27916E+01	6.48101E-01
4.30	1.17535E-01	2.55428E-01	4.60148E-01	1.39549E+01	6.54986E-01
4.40	1.12794E-01	2.46305E-01	4.57942E-01	1.52099E+01	6.61489E-01
4.50	1.08326E-01	2.37624E-01	4.55872E-01	1.65622E+01	6.67635E-01
4.60	1.04112E-01	2.29358E-01	4.53926E-01	1.80178E+01	6.73448E-01
4.70	1.00132E-01	2.21484E-01	4.52096E-01	1.95828E+01	6.78952E-01
4.80	9.63708E-02	2.13980E-01	4.50373E-01	2.12637E+01	6.84167E-01
4.90	9.28123E-02	2.06825E-01	4.48748E-01	2.30671E+01	6.89112E-01
5.00	8.94427E-02	2.00000E-01	4.47214E-01	2.50000E+01	6.93804E-01

Table A.5 Oblique shock wave angle β in degrees for $\gamma = 1.4$

M	θ								
	2°	4°	6°	8°	10°	12°	14°	16°	18°
1.20	61.05								
1.30	53.47	57.42	63.46						
1.40	48.17	51.12	54.63	59.37					
1.50	44.07	46.54	49.33	52.57	56.68	64.36			
1.60	40.72	42.93	45.34	48.03	51.12	54.89	60.54		
1.70	37.93	39.96	42.15	44.53	47.17	50.17	53.77	58.79	
1.80	35.54	37.44	39.48	41.67	44.06	46.69	49.66	53.20	57.99
1.90	33.47	35.28	37.21	39.27	41.49	43.90	46.55	49.54	53.10
2.00	31.65	33.39	35.24	37.21	39.31	41.57	44.03	46.73	49.78
2.10	30.03	31.72	33.51	35.41	37.43	39.59	41.91	44.43	47.21
2.20	28.59	30.24	31.98	33.83	35.78	37.87	40.10	42.49	45.09
2.30	27.29	28.91	30.61	32.42	34.33	36.35	38.51	40.81	43.30
2.40	26.12	27.70	29.38	31.15	33.02	35.01	37.11	39.35	41.75
2.50	25.05	26.61	28.26	30.00	31.85	33.80	35.87	38.06	40.39
2.60	24.07	25.61	27.24	28.97	30.79	32.71	34.75	36.90	39.19
2.70	23.17	24.70	26.31	28.02	29.82	31.73	33.74	35.86	38.11
2.80	22.34	23.85	25.45	27.15	28.94	30.83	32.82	34.92	37.14
2.90	21.58	23.08	24.67	26.35	28.13	30.01	31.99	34.07	36.26
3.00	20.87	22.35	23.94	25.61	27.38	29.25	31.22	33.29	35.47

M	θ							
	20°	22°	24°	26°	28°	30°	32°	34°
1.90	57.90							
2.00	53.42	58.46						
2.10	50.37	54.17	59.77					
2.20	47.98	51.28	55.35	62.69				
2.30	46.01	49.03	52.54	57.08				
2.40	44.34	47.17	50.37	54.18	59.65			
2.50	42.89	45.60	48.60	52.04	56.33			
2.60	41.62	44.24	47.10	50.30	54.09	59.35		
2.70	40.50	43.05	45.81	48.85	52.33	56.69		
2.80	39.49	41.99	44.68	47.60	50.89	54.79	60.43	
2.90	38.58	41.04	43.67	46.51	49.66	53.27	57.93	
3.00	37.76	40.19	42.78	45.55	48.59	52.01	56.18	63.68

(continued)

Table A.5 (continued)

M	θ								
	2°	4°	6°	8°	10°	12°	14°	16°	18°
3.10	20.20	21.68	23.26	24.93	26.69	28.55	30.51	32.57	34.74
3.20	19.59	21.06	22.63	24.29	26.05	27.91	29.86	31.92	34.07
3.30	19.01	20.48	22.04	23.70	25.46	27.31	29.26	31.31	33.46
3.40	18.47	19.93	21.49	23.15	24.90	26.75	28.70	30.75	32.89
3.50	17.96	19.41	20.97	22.63	24.38	26.24	28.18	30.22	32.36
3.60	17.48	18.93	20.49	22.14	23.90	25.75	27.70	29.74	31.88
3.70	17.03	18.48	20.03	21.69	23.44	25.30	27.25	29.29	31.42
3.80	16.60	18.05	19.60	21.26	23.02	24.87	26.82	28.86	31.00
3.90	16.20	17.64	19.20	20.85	22.61	24.47	26.42	28.47	30.60
4.00	15.81	17.26	18.81	20.47	22.23	24.09	26.05	28.10	30.24
4.10	15.45	16.89	18.45	20.11	21.88	23.74	25.70	27.75	29.89
4.20	15.10	16.55	18.10	19.77	21.54	23.41	25.37	27.42	29.56
4.30	14.77	16.22	17.78	19.44	21.22	23.09	25.06	27.11	29.26
4.40	14.46	15.90	17.46	19.13	20.91	22.79	24.76	26.82	28.97
4.50	14.16	15.61	17.17	18.84	20.62	22.51	24.48	26.55	28.70
4.60	13.88	15.32	16.88	18.56	20.35	22.24	24.22	26.29	28.44
4.70	13.60	15.05	16.61	18.30	20.09	21.98	23.97	26.04	28.20
4.80	13.34	14.79	16.36	18.04	19.84	21.74	23.73	25.81	27.97
4.90	13.09	14.54	16.11	17.80	19.60	21.50	23.50	25.59	27.76
5.00	12.85	14.30	15.88	17.57	19.38	21.28	23.29	25.38	27.55

M	θ							
	20°	22°	24°	26°	28°	30°	32°	34°
3.10	37.02	39.42	41.97	44.69	47.65	50.94	54.80	60.20
3.20	36.34	38.72	41.24	43.92	46.81	49.99	53.65	58.35
3.30	35.71	38.08	40.57	43.22	46.06	49.16	52.67	56.96
3.40	35.13	37.49	39.97	42.59	45.39	48.42	51.81	55.84
3.50	34.60	36.95	39.41	42.01	44.77	47.76	51.05	54.89
3.60	34.11	36.45	38.90	41.48	44.21	47.15	50.38	54.07
3.70	33.65	35.99	38.43	40.99	43.70	46.61	49.77	53.34
3.80	33.23	35.56	37.99	40.54	43.23	46.10	49.22	52.70
3.90	32.83	35.16	37.58	40.13	42.80	45.65	48.72	52.13
4.00	32.46	34.79	37.21	39.74	42.40	45.22	48.26	51.61
4.10	32.12	34.44	36.86	39.38	42.03	44.83	47.84	51.13
4.20	31.79	34.11	36.53	39.05	41.69	44.47	47.45	50.70
4.30	31.49	33.81	36.22	38.74	41.37	44.14	47.09	50.30
4.40	31.20	33.53	35.94	38.45	41.07	43.83	46.76	49.94
4.50	30.94	33.26	35.67	38.17	40.79	43.54	46.45	49.60
4.60	30.68	33.01	35.41	37.92	40.53	43.27	46.17	49.29
4.70	30.44	32.77	35.18	37.68	40.28	43.01	45.90	49.00
4.80	30.22	32.54	34.95	37.45	40.05	42.78	45.65	48.73
4.90	30.00	32.33	34.74	37.24	39.84	42.55	45.42	48.48
5.00	29.80	32.13	34.54	37.04	39.64	42.34	45.20	48.24

(continued)

Table A.5 (continued)

M	θ 36°	38°	40°							
3.40	61.92									
3.50	60.09									
3.60	58.79									
3.70	57.76									
3.80	56.89	64.19								
3.90	56.15	62.09								
4.00	55.49	60.83								
4.10	54.92	59.86								
4.20	54.40	59.07								
4.30	53.92	58.40								
4.40	53.50	57.81								
4.50	53.11	57.29	64.34							
4.60	52.74	56.83	63.00							
4.70	52.41	56.40	62.09							
4.80	52.11	56.02	61.37							
4.90	51.82	55.67	60.78							
5.00	51.56	55.35	60.26							

Table A.6 Mach angle and Prandtl-Meyer angle for $\gamma = 1.4$

M	μ	ν	M	μ	ν
1.00	90.00	0.00	1.70	36.03	17.81
1.02	78.64	0.13	1.72	35.55	18.40
1.04	74.06	0.35	1.74	35.08	18.98
1.06	70.63	0.64	1.76	34.62	19.56
1.08	67.81	0.97	1.78	34.18	20.15
1.10	65.38	1.34	1.80	33.75	20.73
1.12	63.23	1.74	1.82	33.33	21.30
1.14	61.31	2.16	1.84	32.92	21.88
1.16	59.55	2.61	1.86	32.52	22.45
1.18	57.94	3.07	1.88	32.13	23.02
1.20	56.44	3.56	1.90	31.76	23.59
1.22	55.05	4.06	1.92	31.39	24.15
1.24	53.75	4.57	1.94	31.03	24.71
1.26	52.53	5.09	1.96	30.68	25.27
1.28	51.38	5.63	1.98	30.33	25.83
1.30	50.28	6.17	2.00	30.00	26.38
1.32	49.25	6.72	2.02	29.67	26.93
1.34	48.27	7.28	2.04	29.35	27.48
1.36	47.33	7.84	2.06	29.04	28.02

(continued)

Table A.6 (continued)

1.38	46.44	8.41	2.08	28.74	28.56
1.40	45.58	8.99	2.10	28.44	29.10
1.42	44.77	9.57	2.12	28.14	29.63
1.44	43.98	10.15	2.14	27.86	30.16
1.46	43.23	10.73	2.16	27.58	30.69
1.48	42.51	11.32	2.18	27.30	31.21
1.50	41.81	11.91	2.20	27.04	31.73
1.52	41.14	12.49	2.22	26.77	32.25
1.54	40.49	13.09	2.24	26.51	32.76
1.56	39.87	13.68	2.26	26.26	33.27
1.58	39.27	14.27	2.28	26.01	33.78
1.60	38.68	14.86	2.30	25.77	34.28
1.62	38.12	15.45	2.32	25.53	34.78
1.64	37.57	16.04	2.34	25.30	35.28
1.66	37.04	16.63	2.36	25.07	35.77
1.68	36.53	17.22	2.38	24.85	36.26
2.40	24.62	36.75	3.10	18.82	51.65
2.42	24.41	37.23	3.12	18.69	52.02
2.44	24.19	37.71	3.14	18.57	52.39
2.46	23.99	38.18	3.16	18.45	52.75
2.48	23.78	38.66	3.18	18.33	53.11
2.50	23.58	39.12	3.20	18.21	53.47
2.52	23.38	39.59	3.22	18.09	53.83
2.54	23.18	40.05	3.24	17.98	54.18
2.56	22.99	40.51	3.26	17.86	54.53
2.58	22.81	40.96	3.28	17.75	54.88
2.60	22.62	41.41	3.30	17.64	55.22
2.62	22.44	41.86	3.32	17.53	55.56
2.64	22.26	42.31	3.34	17.42	55.90
2.66	22.08	42.75	3.36	17.31	56.24
2.68	21.91	43.19	3.38	17.21	56.58
2.70	21.74	43.62	3.40	17.10	56.91
2.72	21.57	44.05	3.42	17.00	57.24
2.74	21.41	44.48	3.44	16.90	57.56
2.76	21.24	44.91	3.46	16.80	57.89
2.78	21.08	45.33	3.48	16.70	58.21
2.80	20.92	45.75	3.50	16.60	58.53
2.82	20.77	46.16	3.52	16.50	58.85
2.84	20.62	46.57	3.54	16.41	59.16
2.86	20.47	46.98	3.56	16.31	59.47
2.88	20.32	47.39	3.58	16.22	59.78
2.90	20.17	47.79	3.60	16.13	60.09
2.92	20.03	48.19	3.62	16.04	60.40
2.94	19.89	48.59	3.64	15.95	60.70

(continued)

Table A.6 (continued)

M	μ	ν	M	μ	ν
2.96	19.75	48.98	3.66	15.86	61.00
2.98	19.61	49.37	3.68	15.77	61.30
3.00	19.47	49.76	3.70	15.68	61.60
3.02	19.34	50.14	3.72	15.59	61.89
3.04	19.20	50.52	3.74	15.51	62.18
3.06	19.07	50.90	3.76	15.42	62.47
3.08	18.95	51.28	3.78	15.34	62.76
3.80	15.26	63.04	4.50	12.84	71.83
3.82	15.18	63.33	4.52	12.78	72.05
3.84	15.09	63.61	4.54	12.72	72.27
3.86	15.01	63.89	4.56	12.67	72.49
3.88	14.94	64.16	4.58	12.61	72.70
3.90	14.86	64.44	4.60	12.56	72.92
3.92	14.78	64.71	4.62	12.50	73.13
3.94	14.70	64.98	4.64	12.45	73.34
3.96	14.63	65.25	4.66	12.39	73.55
3.98	14.55	65.52	4.68	12.34	73.76
4.00	14.48	65.78	4.70	12.28	73.97
4.02	14.40	66.05	4.72	12.23	74.18
4.04	14.33	66.31	4.74	12.18	74.38
4.06	14.26	66.57	4.76	12.13	74.58
4.08	14.19	66.83	4.78	12.08	74.79
4.10	14.12	67.08	4.80	12.02	74.99
4.12	14.05	67.34	4.82	11.97	75.19
4.14	13.98	67.59	4.84	11.92	75.38
4.16	13.91	67.84	4.86	11.87	75.58
4.18	13.84	68.09	4.88	11.82	75.78
4.20	13.77	68.33	4.90	11.78	75.97
4.22	13.71	68.58	4.92	11.73	76.16
4.24	13.64	68.82	4.94	11.68	76.35
4.26	13.58	69.06	4.96	11.63	76.54
4.28	13.51	69.30	4.98	11.58	76.73
4.30	13.45	69.54	5.00	11.54	76.92
4.32	13.38	69.78			
4.34	13.32	70.01			
4.36	13.26	70.24			
4.38	13.20	70.48			
4.40	13.14	70.71			
4.42	13.08	70.93			
4.44	13.02	71.16			
4.46	12.96	71.39			
4.48	12.90	71.61			

Table A.7 Thermodynamic properties of steam, temperature table

T °C	p bar	$v_f \times 10^3$ m³/kg	v_g m³/kg	u_f kJ/kg	u_g kJ/kg	h_f kJ/kg	h_g kJ/kg	s_f kJ/kg K	s_g kJ/kg K
0.01	0.00611	1.0002	206.136	0.00	2375.3	0.01	2501.3	0.0000	9.1562
1	0.00657	1.0002	192.439	4.18	2375.9	4.183	2502.4	0.0153	9.1277
2	0.00706	1.0001	179.762	8.40	2377.3	8.401	2504.2	0.0306	9.1013
3	0.00758	1.0001	168.016	12.61	2378.7	12.61	2506.0	0.0459	9.0752
4	0.00814	1.0001	157.126	16.82	2380.0	16.82	2507.9	0.0611	9.0492
5	0.00873	1.0001	147.024	21.02	2381.4	21.02	2509.7	0.0763	9.0236
6	0.00935	1.0001	137.647	25.22	2382.8	25.22	2511.5	0.0913	8.9981
7	0.01002	1.0001	128.939	29.41	2384.2	29.42	2513.4	0.1063	8.9729
8	0.01073	1.0002	120.847	33.61	2385.6	33.61	2515.2	0.1213	8.9479
9	0.01148	1.0002	113.323	37.80	2386.9	37.80	2517.1	0.1361	8.9232
10	0.01228	1.0003	106.323	41.99	2388.3	41.99	2518.9	0.1510	8.8986
11	0.01313	1.0004	99.808	46.17	2389.7	46.18	2520.7	0.1657	8.8743
12	0.01403	1.0005	93.740	50.36	2391.1	50.36	2522.6	0.1804	8.8502
13	0.01498	1.0006	88.086	54.55	2392.4	54.55	2524.4	0.1951	8.8263
14	0.01599	1.0008	82.814	58.73	2393.8	58.73	2526.2	0.2097	8.8027
15	0.01706	1.0009	77.897	62.92	2395.2	62.92	2528.0	0.2242	8.7792
16	0.01819	1.0011	73.308	67.10	2396.6	67.10	2529.9	0.2387	8.7560
17	0.01938	1.0012	69.023	71.28	2397.9	71.28	2531.7	0.2532	8.7330
18	0.02064	1.0014	65.019	75.47	2399.3	75.47	2533.5	0.2676	8.7101
19	0.02198	1.0016	61.277	79.65	2400.7	79.65	2535.3	0.2819	8.6875
20	0.02339	1.0018	57.778	83.83	2402.0	83.84	2537.2	0.2962	8.6651
21	0.02488	1.0020	54.503	88.02	2403.4	88.02	2539.0	0.3104	8.6428
22	0.02645	1.0023	51.438	92.20	2404.8	92.20	2540.8	0.3246	8.6208
23	0.02810	1.0025	48.568	96.38	2406.1	96.39	2542.6	0.3388	8.5990
24	0.02985	1.0027	45.878	100.57	2407.5	100.57	2544.5	0.3529	8.5773
25	0.03169	1.0030	43.357	104.75	2408.9	104.75	2546.3	0.3670	8.5558
26	0.03363	1.0033	40.992	108.93	2410.2	108.94	2548.1	0.3810	8.5346
27	0.03567	1.0035	38.773	113.12	2411.6	113.12	2549.9	0.3949	8.5135
28	0.03782	1.0038	36.690	117.30	2413.0	117.30	2551.7	0.4088	8.4926
29	0.04008	1.0041	34.734	121.48	2414.3	121.49	2553.5	0.4227	8.4718
30	0.04246	1.0044	32.896	125.67	2415.7	125.67	2555.3	0.4365	8.4513
31	0.04495	1.0047	31.168	129.85	2417.0	129.85	2557.1	0.4503	8.4309
32	0.04758	1.0050	29.543	134.03	2418.4	134.04	2559.0	0.4640	8.4107
33	0.05033	1.0054	28.014	138.22	2419.8	138.22	2560.8	0.4777	8.3906
34	0.05323	1.0057	26.575	142.40	2421.1	142.41	2562.6	0.4914	8.3708
35	0.05627	1.0060	25.220	146.58	2422.5	146.59	2564.4	0.5050	8.3511
40	0.07381	1.0079	19.528	167.50	2429.2	167.50	2573.4	0.5723	8.2550
45	0.09593	1.0099	15.263	188.41	2435.9	188.42	2582.3	0.6385	8.1629
50	0.12345	1.0122	12.037	209.31	2442.6	209.33	2591.2	0.7037	8.0745

(continued)

Table A.7 (continued)

| T | p | $v_f \times 10^3$ | v_g | u_f | u_g | h_f | h_g | s_f | s_g |
°C	bar	m³/kg	m³/kg	kJ/kg	kJ/kg	kJ/kg	kJ/kg	kJ/kg K	kJ/kg K
55	0.1575	1.0146	9.5726	230.22	2449.2	230.24	2600.0	0.7679	7.9896
60	0.1993	1.0171	7.6743	251.13	2455.8	251.15	2608.8	0.8312	7.9080
65	0.2502	1.0199	6.1996	272.05	2462.4	272.08	2617.5	0.8935	7.8295
70	0.3118	1.0228	5.0446	292.98	2468.8	293.01	2626.1	0.9549	7.7540
75	0.3856	1.0258	4.1333	313.92	2475.2	313.96	2634.6	1.0155	7.6813
80	0.4737	1.0290	3.4088	334.88	2481.6	334.93	2643.1	1.0753	7.6112
85	0.5781	1.0324	2.8289	355.86	2487.9	355.92	2651.4	1.1343	7.5436
90	0.7012	1.0359	2.3617	376.86	2494.0	376.93	2659.6	1.1925	7.4784
95	0.8453	1.0396	1.9828	397.89	2500.1	397.98	2667.7	1.2501	7.4154
100	1.013	1.0434	1.6736	418.96	2506.1	419.06	2675.7	1.3069	7.3545
105	1.208	1.0474	1.4200	440.05	2512.1	440.18	2683.6	1.3630	7.2956
110	1.432	1.0515	1.2106	461.19	2517.9	461.34	2691.3	1.4186	7.2386
115	1.690	1.0558	1.0370	482.36	2523.5	482.54	2698.8	1.4735	7.1833
120	1.985	1.0603	0.8922	503.57	2529.1	503.78	2706.2	1.5278	7.1297
125	2.320	1.0649	0.7709	524.82	2534.5	525.07	2713.4	1.5815	7.0777
130	2.700	1.0697	0.6687	546.12	2539.8	546.41	2720.4	1.6346	7.0272
135	3.130	1.0746	0.5824	567.46	2545.0	567.80	2727.2	1.6873	6.9780
140	3.612	1.0797	0.5090	588.85	2550.0	589.24	2733.8	1.7394	6.9302
145	4.153	1.0850	0.4464	610.30	2554.8	610.75	2740.2	1.7910	6.8836
150	4.757	1.0904	0.3929	631.80	2559.5	632.32	2746.4	1.8421	6.8381
160	6.177	1.1019	0.3071	674.97	2568.3	675.65	2758.0	1.9429	6.7503
170	7.915	1.1142	0.2428	718.40	2576.3	719.28	2768.5	2.0421	6.6662
180	10.02	1.1273	0.1940	762.12	2583.4	763.25	2777.8	2.1397	6.5853
190	12.54	1.1414	0.1565	806.17	2589.6	807.60	2785.8	2.2358	6.5071
200	15.55	1.1564	0.1273	850.58	2594.7	852.38	2792.5	2.3308	6.4312
210	19.07	1.1726	0.1044	895.43	2598.7	897.66	2797.7	2.4246	6.3572
220	23.18	1.1900	0.0862	940.75	2601.6	943.51	2801.3	2.5175	6.2847
230	27.95	1.2088	0.0716	986.62	2603.1	990.00	2803.1	2.6097	6.2131
240	33.45	1.2292	0.0597	1033.1	2603.1	1037.2	2803.0	2.7013	6.1423
250	39.74	1.2515	0.0501	1080.4	2601.6	1085.3	2800.7	2.7926	6.0717
260	46.89	1.2758	0.0422	1128.4	2598.4	1134.4	2796.2	2.8838	6.0009
270	55.00	1.3026	0.0356	1177.4	2593.2	1184.6	2789.1	2.9751	5.9293
280	64.13	1.3324	0.0302	1227.5	2585.7	1236.1	2779.2	3.0669	5.8565
290	74.38	1.3658	0.0256	1279.0	2575.7	1289.1	2765.9	3.1595	5.7818
300	85.84	1.4037	0.0217	1332.0	2562.8	1344.1	2748.7	3.2534	5.7042
310	98.61	1.4473	0.0183	1387.0	2546.2	1401.2	2727.0	3.3491	5.6226
320	112.8	1.4984	0.0155	1444.4	2525.2	1461.3	2699.7	3.4476	5.5356
340	145.9	1.6373	0.0108	1569.9	2463.9	1593.8	2621.3	3.6587	5.3345
360	186.6	1.8936	0.0070	1725.6	2352.2	1761.0	2482.0	3.9153	5.0542
374.12	220.9	3.1550	0.0031	2029.6	2029.6	2099.3	2099.3	4.4298	4.4298

Table A.8 Thermodynamic properties of steam, pressure table

p	T	$v_f \times 10^3$	v_g	u_f	u_g	h_f	h_g	s_f	s_g
bar	°C	m³/kg	m³/kg	kJ/kg	kJ/kg	kJ/kg	kJ/kg	kJ/kg K	kJ/kg K
0.06	36.17	1.0065	23.737	151.47	2424.0	151.47	2566.5	0.5208	8.3283
0.08	41.49	1.0085	18.128	173.73	2431.0	173.74	2576.0	0.5921	8.2272
0.10	45.79	1.0103	14.693	191.71	2436.8	191.72	2583.7	0.6489	8.1487
0.12	49.40	1.0119	12.377	206.82	2441.6	206.83	2590.1	0.6960	8.0849
0.16	55.30	1.0147	9.4447	231.47	2449.4	231.49	2600.5	0.7718	7.9846
0.20	60.05	1.0171	7.6591	251.32	2455.7	251.34	2608.9	0.8318	7.9072
0.25	64.95	1.0198	6.2120	271.85	2462.1	271.88	2617.4	0.8929	7.8302
0.30	69.09	1.0222	5.2357	289.15	2467.5	289.18	2624.5	0.9438	7.7676
0.40	75.85	1.0263	3.9983	317.48	2476.1	317.52	2636.1	1.0257	7.6692
0.50	81.31	1.0299	3.2442	340.38	2483.1	340.43	2645.3	1.0908	7.5932
0.60	85.92	1.0330	2.7351	359.73	2488.8	359.79	2652.9	1.1451	7.5314
0.70	89.93	1.0359	2.3676	376.56	2493.8	376.64	2659.5	1.1917	7.4793
0.80	93.48	1.0385	2.0895	391.51	2498.1	391.60	2665.3	1.2327	7.4342
0.90	96.69	1.0409	1.8715	405.00	2502.0	405.09	2670.5	1.2693	7.3946
1.00	99.61	1.0431	1.6958	417.30	2505.5	417.40	2675.1	1.3024	7.3592
2.00	120.2	1.0605	0.8865	504.49	2529.2	504.70	2706.5	1.5301	7.1275
2.50	127.4	1.0672	0.7193	535.12	2537.0	535.39	2716.8	1.6073	7.0531
3.00	133.5	1.0731	0.6063	561.19	2543.4	561.51	2725.2	1.6719	6.9923
3.50	138.9	1.0785	0.5246	584.01	2548.8	584.38	2732.4	1.7276	6.9409
4.00	143.6	1.0835	0.4627	604.38	2553.4	604.81	2738.5	1.7768	6.8963
5.00	151.8	1.0925	0.3751	639.74	2561.1	640.29	2748.6	1.8608	6.8216
6.00	158.8	1.1006	0.3158	669.96	2567.2	670.62	2756.7	1.9313	6.7602
7.00	165.0	1.1079	0.2729	696.49	2572.2	697.27	2763.3	1.9923	6.7081
8.00	170.4	1.1147	0.2405	720.25	2576.5	721.14	2768.9	2.0462	6.6627
9.00	175.4	1.1211	0.2150	741.84	2580.1	742.85	2773.6	2.0947	6.6224
10.00	179.9	1.1272	0.1945	761.67	2583.2	762.80	2777.7	2.1387	6.5861
11.00	184.1	1.1329	0.1775	780.06	2585.9	781.31	2781.2	2.1791	6.5531
12.00	188.0	1.1384	0.1633	797.23	2588.3	798.60	2784.3	2.2165	6.5227
13.00	191.6	1.1437	0.1513	813.37	2590.4	814.85	2787.0	2.2514	6.4946
14.00	195.1	1.1488	0.1408	828.60	2592.2	830.21	2789.4	2.2840	6.4684
15.00	198.3	1.1538	0.1318	843.05	2593.9	844.78	2791.5	2.3148	6.4439
16.00	201.4	1.1586	0.1238	856.81	2595.3	858.66	2793.3	2.3439	6.4208
17.00	204.3	1.1633	0.1167	869.95	2596.5	871.93	2795.0	2.3715	6.3990
18.00	207.1	1.1678	0.1104	882.54	2597.7	884.64	2796.4	2.3978	6.3782
19.00	209.8	1.1723	0.1047	894.63	2598.6	896.86	2797.6	2.4230	6.3585
20.00	212.4	1.1766	0.0996	906.27	2599.5	908.62	2798.7	2.4470	6.3397
25.00	224.0	1.1973	0.0800	958.92	2602.3	961.92	2802.2	2.5543	6.2561
30.00	233.9	1.2165	0.0667	1004.59	2603.2	1008.2	2803.3	2.6453	6.1856
35.00	242.6	1.2348	0.0571	1045.26	2602.9	1049.6	2802.6	2.7250	6.1240

(continued)

Table A.8 (continued)

p	T	$v_f \times 10^3$	v_g	u_f	u_g	h_f	h_g	s_f	s_g
bar	°C	m³/kg	m³/kg	kJ/kg	kJ/kg	kJ/kg	kJ/kg	kJ/kg K	kJ/kg K
40.00	250.4	1.2523	0.0498	1082.18	2601.5	1087.2	2800.6	2.7961	6.0690
45.00	257.5	1.2694	0.0441	1116.14	2599.3	1121.9	2797.6	2.8607	6.0188
50.00	264.0	1.2861	0.0394	1147.74	2596.5	1154.2	2793.7	2.9201	5.9726
55.00	270.0	1.3026	0.0356	1177.39	2593.1	1184.6	2789.1	2.9751	5.9294
60.00	275.6	1.3190	0.0324	1205.42	2589.3	1213.3	2783.9	3.0266	5.8886
65.00	280.9	1.3352	0.0297	1232.06	2584.9	1240.7	2778.1	3.0751	5.8500
70.00	285.9	1.3515	0.0274	1257.52	2580.2	1267.0	2771.8	3.1211	5.8130
75.00	290.6	1.3678	0.0253	1281.96	2575.1	1292.2	2765.0	3.1648	5.7774
80.00	295.0	1.3843	0.0235	1305.51	2569.6	1316.6	2757.8	3.2066	5.7431
85.00	299.3	1.4009	0.0219	1328.27	2563.8	1340.2	2750.1	3.2468	5.7097
90.00	303.4	1.4177	0.0205	1350.36	2557.6	1363.1	2742.0	3.2855	5.6771
95.00	307.3	1.4348	0.0192	1371.84	2551.1	1385.5	2733.4	3.3229	5.6452
100.0	311.0	1.4522	0.0180	1392.79	2544.3	1407.3	2724.5	3.3592	5.6139
105.0	314.6	1.4699	0.0170	1413.27	2537.1	1428.7	2715.1	3.3944	5.5830
110.0	318.1	1.4881	0.0160	1433.34	2529.5	1449.7	2705.4	3.4288	5.5525
115.0	321.5	1.5068	0.0151	1453.06	2521.6	1470.4	2695.1	3.4624	5.5221
120.0	324.7	1.5260	0.0143	1472.47	2513.4	1490.8	2684.5	3.4953	5.4920
125.0	327.9	1.5458	0.0135	1491.61	2504.7	1510.9	2673.4	3.5277	5.4619
130.0	330.9	1.5663	0.0128	1510.55	2495.7	1530.9	2661.8	3.5595	5.4317
135.0	333.8	1.5875	0.0121	1529.31	2486.2	1550.7	2649.7	3.5910	5.4015
140.0	336.7	1.6097	0.0115	1547.94	2476.3	1570.5	2637.1	3.6221	5.3710
145.0	339.5	1.6328	0.0109	1566.49	2465.9	1590.2	2623.9	3.6530	5.3403
150.0	342.2	1.6572	0.0103	1585.01	2455.0	1609.9	2610.0	3.6838	5.3091
155.0	344.8	1.6828	0.0098	1603.55	2443.4	1629.6	2595.5	3.7145	5.2774
160.0	347.4	1.7099	0.0093	1622.17	2431.3	1649.5	2580.2	3.7452	5.2450
165.0	349.9	1.7388	0.0088	1640.92	2418.4	1669.6	2564.1	3.7761	5.2119
170.0	352.3	1.7699	0.0084	1659.89	2404.8	1690.0	2547.1	3.8073	5.1777
175.0	354.7	1.8033	0.0079	1679.18	2390.2	1710.7	2529.0	3.8390	5.1423
180.0	357.0	1.8399	0.0075	1698.88	2374.6	1732.0	2509.7	3.8714	5.1054
185.0	359.3	1.8801	0.0071	1719.17	2357.7	1754.0	2488.9	3.9047	5.0667
190.0	361.5	1.9251	0.0067	1740.22	2339.3	1776.8	2466.3	3.9393	5.0256
195.0	363.7	1.9762	0.0063	1762.34	2319.0	1800.9	2441.4	3.9756	4.9815
200.0	365.8	2.0357	0.0059	1785.94	2296.2	1826.7	2413.7	4.0144	4.9331
205.0	367.9	2.1076	0.0055	1811.76	2269.7	1855.0	2381.6	4.0571	4.8787
210.0	369.9	2.1999	0.0050	1841.25	2237.5	1887.5	2343.0	4.1060	4.8144
215.0	371.8	2.3362	0.0045	1878.57	2193.9	1928.8	2291.0	4.1684	4.7299
220.9	374.1	3.1550	0.0316	2029.60	2029.6	2099.3	2099.3	4.4298	4.4298

Table A.9 Thermodynamic properties of superheated steam

T	v	u	h	s	T	v	u	h	s
°C	m³/kg	kJ/kg	kJ/kg	kJ/kg K	°C	m³/kg	kJ/kg	kJ/kg	kJ/kg K
$p = 0.06$ bar					$p = 0.35$ bar				
36.17	23.739	2424.0	2566.5	8.3283	72.67	4.531	2472.1	2630.7	7.7148
80	27.133	2486.7	2649.5	8.5794	80	4.625	2483.1	2645.0	7.7553
120	30.220	2544.1	2725.5	8.7831	120	5.163	2542.0	2722.7	7.9637
160	33.303	2602.2	2802.0	8.9684	160	5.697	2600.7	2800.1	8.1512
200	36.384	2660.9	2879.2	9.1390	200	6.228	2659.9	2877.9	8.3229
240	39.463	2720.6	2957.4	9.2975	240	6.758	2719.8	2956.3	8.4821
280	42.541	2781.2	3036.4	9.4458	280	7.287	2780.6	3035.6	8.6308
320	45.620	2842.7	3116.4	9.5855	320	7.816	2842.2	3115.8	8.7707
360	48.697	2905.2	3197.4	9.7176	360	8.344	2904.8	3196.9	8.9031
400	51.775	2968.8	3279.5	9.8433	400	8.872	2968.5	3279.0	9.0288
440	54.852	3033.4	3362.6	9.9632	440	9.400	3033.2	3362.2	9.1488
500	59.468	3132.4	3489.2	10.134	500	10.192	3132.2	3488.9	9.3194
$p = 0.7$ bar					$p = 1$ bar				
89.93	2.368	2493.8	2659.5	7.4793	99.61	1.6958	2505.5	2675.1	7.3592
120	2.5709	2539.3	2719.3	7.6370	120	1.7931	2537.0	2716.3	7.4665
160	2.8407	2599.0	2797.8	7.8272	160	1.9838	2597.5	2795.8	7.6591
200	3.1082	2658.7	2876.2	8.0004	200	2.1723	2657.6	2874.8	7.8335
240	3.3744	2718.9	2955.1	8.1603	240	2.3594	2718.1	2954.0	7.9942
280	3.6399	2779.8	3034.6	8.3096	280	2.5458	2779.2	3033.8	8.1438
320	3.9049	2841.6	3115.0	8.4498	320	2.7317	2841.1	3114.3	8.2844
360	4.1697	2904.4	3196.2	8.5824	360	2.9173	2904.0	3195.7	8.4171
400	4.4342	2968.1	3278.5	8.7083	400	3.1027	2967.7	3278.0	8.5432
440	4.6985	3032.8	3361.7	8.8285	440	3.2879	3032.5	3361.3	8.6634
480	4.9627	3098.6	3446.0	8.9434	480	3.4730	3098.3	3445.6	8.7785
520	5.2269	3165.4	3531.3	9.0538	520	3.6581	3165.2	3531.0	8.8889
$p = 1.5$ bar					$p = 3$ bar				
111.37	1.1600	2519.3	2693.3	7.2234	133.55	0.6060	2543.4	2725.2	6.9923
160	1.3174	2594.9	2792.5	7.4660	160	0.6506	2586.9	2782.1	7.1274
200	1.4443	2655.8	2872.4	7.6425	200	0.7163	2650.2	2865.1	7.3108
240	1.5699	2716.7	2952.2	7.8044	240	0.7804	2712.6	2946.7	7.4765
280	1.6948	2778.2	3032.4	7.9548	280	0.8438	2775.0	3028.1	7.6292
320	1.8192	2840.3	3113.2	8.0958	320	0.9067	2837.8	3109.8	7.7716
360	1.9433	2903.3	3194.8	8.2289	360	0.9692	2901.2	3191.9	7.9057
400	2.0671	2967.2	3277.2	8.3552	400	1.0315	2965.4	3274.9	8.0327
440	2.1908	3032.0	3360.6	8.4756	440	1.0937	3030.5	3358.7	8.1536
480	2.3144	3097.9	3445.1	8.5908	480	1.1557	3096.6	3443.4	8.2692
520	2.4379	3164.8	3530.5	8.7013	520	1.2177	3163.7	3529.0	8.3800
560	2.5613	3232.9	3617.0	8.8077	560	1.2796	3231.9	3615.7	8.4867

(continued)

Table A.9 (continued)

T	v	u	h	s	T	v	u	h	s
°C	m³/kg	kJ/kg	kJ/kg	kJ/kg K	°C	m³/kg	kJ/kg	kJ/kg	kJ/kg K
p = 5 bar					p = 7 bar				
151.86	0.3751	2561.1	2748.6	6.8216	164.97	0.2729	2572.2	2763.3	6.7081
180	0.4045	2609.5	2811.7	6.9652	180	0.2846	2599.6	2798.8	6.7876
220	0.4449	2674.9	2897.4	7.1463	220	0.3146	2668.1	2888.4	6.9771
260	0.4840	2738.9	2980.9	7.3092	260	0.3434	2733.9	2974.2	7.1445
300	0.5225	2802.5	3063.7	7.4591	300	0.3714	2798.6	3058.5	7.2970
340	0.5606	2866.3	3146.6	7.5989	340	0.3989	2863.2	3142.4	7.4385
380	0.5985	2930.7	3229.9	7.7304	380	0.4262	2928.1	3226.4	7.5712
420	0.6361	2995.7	3313.8	7.8550	420	0.4533	2993.6	3310.9	7.6966
460	0.6736	3061.6	3398.4	7.9738	460	0.4802	3059.8	3395.9	7.8160
500	0.7109	3128.5	3483.9	8.0873	500	0.5070	3126.9	3481.8	7.9300
540	0.7482	3196.3	3570.4	8.1964	540	0.5338	3194.9	3568.5	8.0393
560	0.7669	3230.6	3614.0	8.2493	560	0.5741	3229.2	3612.2	8.09243
p = 10 bar					p = 15 bar				
179.91	0.1945	2583.2	2777.7	6.5861	198.32	0.1318	2593.9	2791.5	6.4439
220	0.2169	2657.5	2874.3	6.7904	220	0.1405	2638.1	2849.0	6.5630
260	0.2378	2726.1	2963.9	6.9652	260	0.1556	2712.6	2945.9	6.7521
300	0.2579	2792.7	3050.6	7.1219	300	0.1696	2782.5	3036.9	6.9168
340	0.2776	2858.5	3136.1	7.2661	340	0.1832	2850.4	3125.2	7.0657
380	0.2970	2924.2	3221.2	7.4006	380	0.1965	2917.6	3212.3	7.2033
420	0.3161	2990.3	3306.5	7.5273	420	0.2095	2984.8	3299.1	7.3322
460	0.3352	3057.0	3392.2	7.6475	460	0.2224	3052.3	3385.9	7.4540
500	0.3541	3124.5	3478.6	7.7622	500	0.2351	3120.4	3473.1	7.5699
540	0.3729	3192.8	3565.7	7.8721	540	0.2478	3189.2	3561.0	7.6806
580	0.3917	3262.0	3653.7	7.9778	580	0.2605	3258.8	3649.6	7.7870
620	0.4105	3332.2	3742.7	8.0797	620	0.2730	3329.4	3739.0	7.8894
p = 20 bar					p = 30 bar				
212.42	0.0996	2599.5	2798.7	6.3397	233.90	0.06667	2603.2	2803.3	6.1856
240	0.1084	2658.8	2875.6	6.4937	240	0.06818	2618.9	2823.5	6.2251
280	0.1200	2735.6	2975.6	6.6814	280	0.07710	2709.0	2940.3	6.4445
320	0.1308	2807.3	3068.8	6.8441	320	0.08498	2787.6	3042.6	6.6232
360	0.1411	2876.7	3158.9	6.9911	360	0.09232	2861.3	3138.3	6.7794
400	0.1512	2945.1	3247.5	7.1269	400	0.09935	2932.7	3230.7	6.9210
440	0.1611	3013.4	3335.6	7.2539	440	0.10618	3003.0	3321.5	7.0521
480	0.1708	3081.9	3423.6	7.3740	480	0.11287	3073.0	3411.6	7.1750
520	0.1805	3150.9	3511.9	7.4882	520	0.11946	3143.2	3501.6	7.2913
560	0.1901	3220.6	3600.7	7.5975	560	0.12597	3213.8	3591.7	7.4022
600	0.1996	3291.0	3690.2	7.7024	600	0.13243	3285.0	3682.3	7.5084
640	0.2091	3362.4	3780.5	7.8036	640	0.13884	3357.0	3773.5	7.6105

(continued)

Table A.9 (continued)

T	v	u	h	s	T	v	u	h	s
°C	m³/kg	kJ/kg	kJ/kg	kJ/kg K	°C	m³/kg	kJ/kg	kJ/kg	kJ/kg K
$p = 40$ bar					$p = 60$ bar				
250.38	0.04978	2601.5	2800.6	6.0690	276.62	0.03244	2589.3	2783.9	5.8886
280	0.05544	2679.0	2900.8	6.2552	280	0.03317	2604.7	2803.7	5.9245
320	0.06198	2766.6	3014.5	6.4538	320	0.03874	2719.0	2951.5	6.1830
360	0.06787	2845.3	3116.7	6.6207	360	0.04330	2810.6	3070.4	6.3771
400	0.07340	2919.8	3213.4	6.7688	400	0.04739	2892.7	3177.0	6.5404
440	0.07872	2992.3	3307.2	6.9041	440	0.05121	2970.2	3277.4	6.6854
480	0.08388	3064.0	3399.5	7.0301	480	0.05487	3045.3	3374.5	6.8179
520	0.08894	3135.4	3491.1	7.1486	520	0.05840	3119.4	3469.8	6.9411
560	0.09392	3206.9	3582.6	7.2612	560	0.06186	3193.0	3564.1	7.0571
600	0.09884	3278.9	3674.3	7.3687	600	0.06525	3266.6	3658.1	7.1673
640	0.10372	3351.5	3766.4	7.4718	640	0.06859	3340.5	3752.1	7.2725
680	0.10855	3424.9	3859.1	7.5711	680	0.07189	3414.9	3846.3	7.3736
$p = 80$ bar					$p = 100$ bar				
295.04	0.02352	2569.6	2757.8	5.7431	311.04	0.01802	2544.3	2724.5	5.6139
320	0.02681	2661.7	2876.2	5.9473	320	0.01925	2588.2	2780.6	5.7093
360	0.03088	2771.9	3018.9	6.1805	360	0.02330	2728.0	2961.0	6.0043
400	0.03431	2863.5	3138.0	6.3630	400	0.02641	2832.0	3096.1	6.2114
440	0.03742	2946.8	3246.2	6.5192	440	0.02911	2922.3	3213.4	6.3807
480	0.04034	3026.0	3348.6	6.6589	480	0.03160	3005.8	3321.8	6.5287
520	0.04312	3102.9	3447.8	6.7873	520	0.03394	3085.9	3425.3	6.6625
560	0.04582	3178.6	3545.2	6.9070	560	0.03619	3164.0	3525.8	6.7862
600	0.04845	3254.0	3641.5	7.0200	600	0.03836	3241.1	3624.7	6.9022
640	0.05102	3329.3	3737.5	7.1274	640	0.04048	3317.9	3722.7	7.0119
680	0.05356	3404.9	3833.4	7.2302	680	0.04256	3394.6	3820.3	7.1165
720	0.05607	3480.9	3929.4	7.3289	720	0.04461	3471.6	3917.7	7.2167
$p = 120$ bar					$p = 140$ bar				
324.75	0.01426	2513.4	2684.5	5.4920	336.75	0.01148	2476.3	2637.1	5.3710
360	0.01810	2677.1	2894.4	5.8341	360	0.01421	2616.0	2815.0	5.6579
400	0.02108	2797.8	3050.7	6.0739	400	0.01722	2760.2	3001.3	5.9438
440	0.02355	2896.3	3178.9	6.2589	440	0.01955	2868.8	3142.5	6.1477
480	0.02576	2984.9	3294.0	6.4161	480	0.02157	2963.1	3265.2	6.3152
520	0.02781	3068.4	3402.1	6.5559	520	0.02343	3050.3	3378.3	6.4616
560	0.02976	3149.0	3506.1	6.6839	560	0.02517	3133.6	3485.9	6.5940
600	0.03163	3228.0	3607.6	6.8029	600	0.02683	3214.7	3590.3	6.7163
640	0.03345	3306.3	3707.7	6.9150	640	0.02843	3294.5	3692.5	6.8309
680	0.03523	3384.3	3807.0	7.0214	680	0.02999	3373.8	3793.6	6.9392
720	0.03697	3462.3	3906.0	7.1231	720	0.03152	3452.8	3894.1	7.0425
760	0.03869	3540.6	4004.8	7.2207	760	0.03301	3532.0	3994.2	7.1413

(continued)

Table A.9 (continued)

T	v	u	h	s	T	v	u	h	s
°C	m³/kg	kJ/kg	kJ/kg	kJ/kg K	°C	m³/kg	kJ/kg	kJ/kg	kJ/kg K
$p = 160$ bar					$p = 180$ bar				
347.44	0.00931	2431.3	2580.2	5.2450	357.06	0.00750	2374.6	2509.7	5.1054
360	0.01105	2537.5	2714.3	5.4591	360	0.00810	2418.3	2564.1	5.1916
400	0.01427	2718.5	2946.8	5.8162	400	0.01191	2671.7	2886.0	5.6872
440	0.01652	2839.6	3104.0	6.0433	440	0.01415	2808.5	3063.2	5.9432
480	0.01842	2940.5	3235.3	6.2226	480	0.01596	2916.9	3204.2	6.1358
520	0.02013	3031.8	3353.9	6.3761	520	0.01756	3012.7	3328.8	6.2971
560	0.02172	3117.9	3465.4	6.5133	560	0.01903	3101.9	3444.5	6.4394
600	0.02322	3201.1	3572.6	6.6390	600	0.02041	3187.3	3554.8	6.5687
640	0.02466	3282.6	3677.2	6.7561	640	0.02173	3270.5	3661.7	6.6885
680	0.02606	3363.1	3780.1	6.8664	680	0.02301	3352.4	3766.5	6.8008
720	0.02742	3443.3	3882.1	6.9712	720	0.02424	3433.7	3870.0	6.9072
760	0.02876	3523.4	3983.5	7.0714	760	0.02545	3514.7	3972.8	7.0086
$p = 200$ bar					$p = 240$ bar				
365.81	0.00588	2296.2	2413.7	4.9331					
400	0.00995	2617.9	2816.9	5.5521	400	0.00673	2476.0	2637.5	5.2365
440	0.01223	2775.2	3019.8	5.8455	440	0.00929	2700.9	2923.9	5.6511
480	0.01399	2892.3	3172.0	6.0534	480	0.01100	2839.9	3103.9	5.8971
520	0.01551	2993.1	3303.2	6.2232	520	0.01241	2952.1	3250.0	6.0861
560	0.01688	3085.5	3423.2	6.3708	560	0.01366	3051.8	3379.5	6.2456
600	0.01817	3173.3	3536.7	6.5039	600	0.01480	3144.6	3499.8	6.3866
640	0.01939	3258.2	3646.0	6.6264	640	0.01587	3233.3	3614.3	6.5148
680	0.02056	3341.6	3752.8	6.7408	680	0.01690	3319.6	3725.1	6.6336
720	0.02170	3424.0	3857.9	6.8488	720	0.01788	3404.3	3833.4	6.7450
760	0.02280	3505.9	3961.9	6.9515	760	0.01883	3488.2	3940.2	6.8504
800	0.02388	3587.8	4065.4	7.0498	800	0.01976	3571.7	4046.0	6.9508
$p = 280$ bar					$p = 320$ bar				
400	0.00383	2221.7	2328.8	4.7465	400	0.00237	1981.0	2056.8	4.3252
440	0.00712	2613.5	2812.9	5.4497	440	0.00543	2509.0	2682.9	5.2325
480	0.00885	2782.7	3030.5	5.7472	480	0.00722	2720.5	2951.5	5.5998
520	0.01019	2908.9	3194.3	5.9592	520	0.00853	2863.4	3136.2	5.8390
560	0.01135	3016.8	3334.6	6.1319	560	0.00962	2980.6	3288.4	6.0263
600	0.01239	3115.1	3462.1	6.2815	600	0.01059	3084.9	3423.8	6.1851
640	0.01336	3207.9	3582.0	6.4158	640	0.01148	3182.0	3549.4	6.3258
680	0.01428	3297.2	3697.0	6.5390	680	0.01232	3274.6	3668.8	6.4538
720	0.01516	3384.4	3808.8	6.6539	720	0.01312	3364.3	3784.0	6.5722
760	0.01600	3470.3	3918.4	6.7621	760	0.01388	3452.3	3896.4	6.6832
800	0.01682	3555.5	4026.5	6.8647	800	0.01462	3539.2	4006.9	6.7881

Index

© The Author(s) 2021
V. Babu, *Fundamentals of Gas Dynamics*,
https://doi.org/10.1007/978-3-030-60819-4